DISCARD

REMEMBERING
OUR
CHILDHOOD

REMEMBERING OUR CHILDHOOD

How Memory Betrays Us

....

BY KARL SABBAGH

OXFORD
UNIVERSITY PRESS

OXFORD
UNIVERSITY PRESS

Great Clarendon Street, Oxford OX2 6DP

Oxford University Press is a department of the University of Oxford.
It furthers the University's objective of excellence in research, scholarship,
and education by publishing worldwide in

Oxford New York

Auckland Cape Town Dar es Salaam Hong Kong Karachi
Kuala Lumpur Madrid Melbourne Mexico City Nairobi
New Delhi Shanghai Taipei Toronto

With offices in

Argentina Austria Brazil Chile Czech Republic France Greece
Guatemala Hungary Italy Japan Poland Portugal Singapore
South Korea Switzerland Thailand Turkey Ukraine Vietnam

Oxford is a registered trade mark of Oxford University Press
in the UK and in certain other countries

Published in the United States
by Oxford University Press Inc., New York

© Karl Sabbagh 2009

British Library Cataloguing in Publication Data

Data available

Library of Congress Cataloging in Publication Data

Data available

Typeset by SPI Publisher Services, Pondicherry, India
Printed in Great Britain
on acid-free paper by
Clays Ltd, St. Ives plc

ISBN 978–0–19–921840–0

1 3 5 7 9 10 8 6 4 2

Contents

Acknowledgements

I was encouraged to write this book by Professor Larry Weiskrantz, Emeritus Professor of Psychology at the University of Oxford. He was also responsible for my first contact with the world of experimental psychology when he was my undergraduate supervisor at Cambridge in the 1960s. He has been very helpful in reading a draft of the manuscript and making comments. I have also valued the advice of Professor Elizabeth Loftus at the University of California at Irvine, who, as a leading researcher in the field, has been responsible for key discoveries in the 1980s and 1990s which have transformed our ideas of the nature of memory. She, too, has provided very helpful comments and research suggestions during the writing of this book, and invited me to spend a day with her in her department at UCLA where we discussed many of the topics.

Six other psychologists working in the field of memory were generous with their time during my research phase. When I quote from these psychologists in the book without a specific endnote reference, these quotes are taken from interviews I recorded and transcribed on the dates given below:

> Madeline Eacott, 6 February 2007
> Robyn Fivush, 2 April 2007
> Jennifer Freyd, 11 October 2006
> Gail Goodman, 18 October 2006
> Elizabeth Loftus, 16 October 2006
> Richard McNally, 14 September 2006
> Jonathan Schooler, 12 October 2006

I would like to thank a number of friends and acquaintances for supplying me with their childhood memories for Chapter 1 of this book.

They include the following: Roy Ackerman, Vic Allen, Patrick Bateson, Shelley Baxter, Grit Bendixen, Michael Berry, Fergus Bordewich, Janet Browne, Rob Buckman, Marcus Clapham, Alyce-Faye Eichelberger Cleese, John Cleese, Andy Cottom, Lorayne Crawford, Richard Dawkins, Daniel Dennett, Adrian Dicks, Richard Eyre, Harley Fine, Sally Fryer, Graeme Garden, Howard Graff, Richard Gregory, Christopher Hale, Carey Harrison, Nicholas Humphrey, Katie Jarvis, David Kennard, Donna Knight, Barry Kushner, Harvey Marcovitch, Ariel Marguin, Marshall Marinker, Elena Maroko, Callum McCarthy, Penny McCarthy, Jonathan Miller, Josceline Newell, Richard Newell, Tony Palmer, Ilan Pappé, William Percy, Roger Pringle, Matt Ridley, Bill Rosen, Sue Sabbagh, Jonathan Sale, Angela Scholar, Michael Scholar, John Shrapnel, Christopher Sykes, Nicholas Wade, Terry Waite, Fenella Warden, Andrew Watson, Christabel Watson, Peter Wilson.

I would also like to thank John Montagu, the Earl of Sandwich, for allowing me to share his experience of revisiting his childhood home for the first time, and Dr Janet Browne, for drawing my attention to Charles Darwin's earliest memory.

I am particularly indebted to the efforts by Mark Pendergrast to gather evidence of prevailing views among therapists, 'survivors', and accused parents in the Recovered Memory movement, and I have drawn heavily on the transcripts of interviews included in his book, *Victims of Memory*. I have used the out of print HarperCollins UK edition, but the US edition* is still available through www.upperaccess.com.

I am grateful for the permission of Mark Prendergast and his publisher, Steve Carlson, to quote from *Victims of Memory* and I would also like to thank Susan Clancy for permission to quote from her book *Abducted: How People Come to Believe They Were Kidnapped by Aliens*, published by Harvard University Press, Cambridge, Massachusetts, 2005.

My researches for this book were partly funded by a grant from the Authors Foundation, administered by the Society of Authors, whose generosity I would like to acknowledge.

I have appreciated working with Latha Menon, my editor at OUP, and valued her unobtrusive advice and support.

* *Victims of Memory: Sex abuse accusations and shattered lives*, Second edition (Upper Access, Hinesburg, Vt., 1996) ISBN 978–0–942679–18–2.

ACKNOWLEDGEMENTS

Finally, I would like to thank Madeline Greenhalgh, Director of the British False Memory Society, and its Scientific Advisory Board, whose meetings I have attended as an independent member. Their individual professional specialisms have been very helpful in clarifying the scientific issues in the field of memory, but I should also like to make clear that the views expressed in this book are my own and nobody else's.

1

· · · · · · · · ·

'To remember for years'

When I was a child, my mother would recite one of her favourite poems. I remember it like this:

> Three ducks on a pond,
> And the green grass beyond.
> What a thing to remember for years,
> To remember with tears.

Recently, I was looking through a commonplace book which I found in a second-hand bookshop. It was handwritten, and there was a poem or passage for each day. On the page for 19 April was written:

> Four ducks on a pond,
> A grass bank beyond,
> A blue sky of spring,
> White clouds on the wing:
> What a little thing
> To remember for years,
> To remember with tears!

Not three ducks, four. Not green grass but a green bank. And two or three extra lines (which actually, for me, made it a less satisfying verse. My version, or my mother's, had more of a haiku-like quality to it).

This experience raised in a small way the pitfalls of memory. The poet, William Allingham, wrote his poem, version A. My mother read that and memorized it, version B (which may have been identical with

A, but it may not). Years later, she then recited C to me (which may have been identical to B, or may not), which I now remember as D (which may be identical to C, or may not).

Between each 'recitation', time, context, interference, and new associations had intervened in ways which could have changed the remembered verses, and we have no way of knowing whether during the period after my mother's first memory the words changed, or whether the change took place in my own memory over the last fifty or so years. There's even one further question one can ask. Were there *really* four ducks on the pond, that the poet remembered for years, in tears, or between seeing them and writing about it did he forget to include that brown mallard, tucked under the reeds on the far side and camouflaged by the mottled shadows?

'Remembering for years' is something which is familiar to most of us. Many people respond with enthusiasm when asked to describe their earliest memory, and most people will have one. Of course, literally, everyone must have *an* earliest memory, but some deny having a memory from early childhood. (Although one friend, having denied that he could remember anything in his early childhood then gave me three very interesting mental snapshots of childhood events.)

This is a topic of more than academic interest (although there's nothing wrong with academic interest—most 'useful' discoveries of science have begun with intellectual curiosity rather than with some practical aim.) Childhood memories can lead to adult harm, if they are taken to be accurate replications of real events.

Most childhood memories are benign. Even if they relate to moments of unhappiness rather than pleasure, as adults we can look back from a distance of decades and put them into context as the trivial events they were. Important at the time but overtaken by the weightier pleasures and pains of adulthood.

But over the last twenty years psychologists, psychotherapists, and lawyers have found themselves dealing with childhood memories which are much less benign, and this has driven the exploration of scientific issues which had been little studied before.

Here are a few examples from the last twenty years of the way childhood memories—and in particular the issue of accuracy—have come to assume much more importance:

> The daughter of a retired fireman, when an adult, remembers as a child seeing her father murder her eight-year-old schoolfriend.

> A young adult remembers that her father raped her numerous times between the ages of five and sixteen, and forced her to have sex with the family dog.

> In a nursery school in California, several children remember a series of abuses being carried out by the teachers. They include being photographed while performing nude somersaults; playing a nude version of 'Cowboys and Indians', sometimes with the Indians sexually assaulting the cowboys, and sometimes vice versa; sexual assaults taking place on farms, in circus houses, in the homes of strangers, in car washes, in store rooms, and in a 'secret room' accessible by a tunnel; watching animal sacrifices performed by teachers wearing robes and masks in a candle-lit ceremony at St Cross Episcopal Church; being taken to a cemetery where the children were forced to use pickaxes and shovels to dig up coffins.

> A young adult remembers incidents when her father made her masturbate him regularly from the age of three.

> A young woman remembers her father raping her when she was small, and beating to death a six-year-old girl in front of her.

> The American television actor, Roseanne Barr, reports detailed memories of her mother molesting her in her crib when she was six months old.

> A middle-aged woman remembers her father and his gay lover sexually abusing her when she was between three and six months old.

Because such accounts involve what would be criminal acts if they are true, the need to understand the accuracy of childhood memory has led psychologists to investigate the processes that underlie our memories, both as children and as adults. And even childhood memories of abuse that do not end up offered as evidence in court raise issues of truth, falsehood, and accuracy when they become believed by 'victims', their relatives, and, as we shall see, their psychotherapists.

But let us start by considering the more 'normal' type of account that most of us can produce when asked what we remember of our early childhood. In an informal survey of friends and acquaintances for this book, I gathered sixty or so 'earliest memories' to raise questions about what we remember of our childhoods and what such memories tell us about the psychological processes of memory.*

Collecting spontaneous memories of their childhood from adults immediately raises issue of truth and accuracy. Elizabeth Loftus, one of the leading researchers in the field of memory, told me one of her early memories and then added a caveat:

For much of my life I thought my earliest memory was going to see *The Greatest Show on Earth*. I remember being very happy, even saying 'this is the happiest day of my life.' I thought I was around four or five years old. Decades later, after I became a memory scientist, I came across an encyclopaedia of movies. I looked up *The Greatest Show on Earth* and was shocked to find the movie came out when I was eight years old. So it wasn't my earliest memory, since I had memories for things that happened before age eight.

People sometimes take pride in how young they were at the time of their earliest memories. But very early memories raise scepticism in some psychologists, who have discovered a phenomenon called childhood amnesia, which relates to the first two or three years of life. One leading memory researcher, Richard McNally, a professor of psychology at Harvard, has written:

Most adults are incapable of recalling any experiences they had before the age of 3 or 4. The cognitive psychologist David Rubin found that about 90 percent of the childhood episodes recalled by adults occurred after their fourth birthday, and many adults remember hardly anything from before the age of 7. Earliest memories usually contain visual imagery, and are often little more than fragmentary 'snapshots' devoid of narrative structure.[1]

If you take people's childhood memories at face value—at least in my sample—this is just plain wrong. Here is a selection of those

---

* The full list of memories is given in the Appendix.

memories, in the rememberers' own words, that claim to derive from ages below three (claimed ages given in brackets[*]).

- Sunlight sparkling through leaves. (Age 1)
- Being in some kind of pram and the yellow flowering branches over my head and it was warm and sunny and I was happy. (Age 1.5)
- I remember the look on my mother's face as she grabbed me up and ran with me downstairs. I remember the house shaking and bottles falling around us. (Age 1.9)
- Car stopped, getting out at odd grassy place, teddy or doll needs to get out too. (Age 1.9)
- Waking up in cot, lying, seeing yellow curtains waving in breeze, somebody nearby having a shave. Wallpaper with dolls. (Age 1.9)
- Seeing a very dark man at our gate. I was standing by a high wooden gate. The gate opened and I saw the darkest man I've ever seen: skin like bark, thick black stubble, large black eyes. When he spoke I couldn't understand him, and I ran to my mother. She gave him something. Money perhaps. He went. I cried. He was an Italian ex-prisoner of war who was looking for work. (Age 2)
- Playing with dog and cat in farmyard. (Age 2)
- My earliest memory is of me standing on my Grandpa Alf's knees, whilst he was seated. He had an almost completely bald head that was smooth and shiny and I remember kneading his head perhaps because it was such a strange thing for a young child to behold. (My other relatives had hair!). (Age 2)
- Rocking-horse and moving pedal-operated toy horse in large playroom. (Age 2)
- On a horse I am very tall, it is wonderful. (Age 2)
- My brother walked me down our terraced street in Liverpool. I held his hand. We crossed over the big road, which was usually busy and full of cars. We were standing for ages, just waiting. Then... there were two more motorbikes with policemen on and right behind was a black car. I could see the people sitting in the back facing each other like they would on a bus. The back window of the car was down and one of the men had his head out of it a bit and was waving. He had black hair and it was bushy. I waved back. [This took place in Liverpool, and the car was carrying the Beatles.] (Age 2)

* Ages are approximate and expressed in decimals, i.e. 2.5 = 2 years 6 months.

- I was in my pram and I woke up looking at the sun making a line of shadow on the houses opposite us on Rosecroft Avenue. I do remember very clearly the line of the shadow and the sort of beigey look of the walls (they were pebbledash I found out many years later). BUT the really odd thing is that I remember thinking (as a thought-shape not as words of course) 'Here I am again'. So I must have known I'd been there before. (Age 2)
- Holding a woman's hand entering an apartment, walking in the narrow entryway and seeing a bathtub on the left. (Age 2)
- Lying looking up at the sky. (Age 2)
- Standing and looking out and down at a sunlit garden and calling out 'Cecil' to a man I could see below. (Age 2)
- Lying down and looking up at a blue sky with a terracotta red tiled roof and hearing and seeing a piston-engined aircraft flying slowly overhead. (Age 2)
- Being taken to the doctor in Nairobi for an injection. His name was Dr Trim. We had to mount a flight of stairs, then turn right at the top of the stairs through a door to see the doctor. (Age 2)

What are we to make of these 'memories' that describe events before McNally's cut-off point of three or four?

First, McNally and others are interested principally in memories that have 'narrative structure'. The content of many of the earliest memories that my informal sample provided is solely visual imagery— 'Seeing a very dark man', 'rocking horse', 'shadow on a wall', 'leaves against the sky'. However, there *is* also narrative structure—the two-year-old who remembers being taken to see what he later realized were the Beatles, the visit to the doctor in Nairobi.

One of the things that is becoming clearer from the work of researchers who have looked in more detail at the earliest memories is that there seem to be gradations of content, so that the very earliest memories have purely visual—or occasionally aural—content, and that as the child gets older a kind of narrative creeps in, and memories are about episodes rather than just sensory impressions. We can see the evolution from sensory to narrative in the following selection of memories of specific activities ascribed by the rememberers to a period when they were between two and four:

- Lying on my back looking up at the sky with the leaves moving. (Age 2.5)
- I have a clear memory of being taken through a gap in the barbed wire on some sand dunes to get to the beach in Dorset. This little island of memory largely features the sand dunes, the marram grass, and the barbed wire with its gap. (Age 2.5)
- A giant hole in the ground, with lamp-posts at crazy angles surrounding it, and snow on the ground. (Age 2.5)
- Playing on a grassy bank. (Age 2.5)
- Loud bang, confusion everywhere, looking out of upstairs window. (Age 2.5)
- A game: being thrown through the air by large men in a big room—a sweet smell from them. Laughter. (Age 2.75)
- Both parents swimming out to sea, becoming invisible from the sandcastle on which I was standing with my brother and sister, and fearing that parents would not be coming back. Comforted by sister. (Age 2.8)
- I am in a bathroom (large, bright, perhaps sumptuous). I have shat my pants, and am trying without much success to clean them in a sink. Feelings of fear at impending punishment. (Age 3)
- Camp beds with grey blankets on which we had to 'lie down and rest' during first few days in school. (Age 3)
- Playing with my grandfather's porcelain eye before he put it back in place. (Age 3)
- Riding a tricycle in Turner Close. (Age 3)
- I was standing on the balcony of my parents' flat in Park Crescent, watching soldiers marching by. I remember drawing my parents' attention to the way the men were swinging their arms. (Age 3)
- A red brick wall, topped with a white corbel, rough to the touch. In back of the wall, a window, and behind the window, my mother. The wall was at least five feet high, so I must have been held by my father, and placed my hands on the surface of the wall, whose texture remains the most vivid portion of the memory. (Age 3.5)
- I am looking down a grating at a golliwog which has dropped down somehow. I am sad. The sun is shining and has a face like the lemon in the 'Idris when I's dry' advertisement for a brand of lemon squash. This advertisement features in another memory which I think is based on what I was told by my mother, which is that when I saw this advertisement on a bus, I said 'Funny lemon a-crying.' I seem to remember that I found the face rather frightening. (Age 3)

7

Memories from a slightly older age can show some kind of emotional content and sometimes an attempt at interpretation overlaid on the core sensations, the sights and sounds. Often this is presented as if it is what was running through the child's mind at the time although it could well be a later addition, a part of the re-remembering process we all experience.

- The first time I met my Father. I stood with my Mother and Grandparents in the little hallway of my Grandparents' house as a big smiling man in uniform came up the steps through the front door to scoop me into his arms. My father went abroad serving in the RAMC shortly before I was born in 1943. The night (or a few nights) before his return I had a dream which for some reason made a huge impression on me. I was standing at the edge of my Grandparents' lawn looking down at neat rows of cabbages. The cabbages looked like vegetable faces, and they laughed as I played on the lawn. When my father and I met and he picked me up I think I felt confused, and perhaps alarmed. Although I have no distinct memory of saying it, I have often been told that my first words to my father were 'The cabbages laughed at me.' (Age 3)

- Leaning over the parapet of a bridge across railway tracks. My father is holding me up to watch trains. I am alarmed to be in his care and away from my mother, and dislike the smelly clouds of steam that envelop us as each train roars under the bridge. (Age 3)

- Hiding, afraid, under the kitchen table in my grandmother's house in Merthyr Tydfil when an aeroplane went overhead. (Age 3.3)

- Sitting (trapped?) in a wooden high chair, fronted with a little table top, with an insect buzzing around me, strong visual memory of a woman flapping it away, possibly with a cloth. I have another early memory, with no idea of which is first, of being in some spectator stands with my family and seeing a procession, horses, soldiers, coaches, etc. This is associated with a vague recollection of my younger sister wanting to pee and my mother whipping out a plastic pot. (Age 3.3)

- Running between two people, whom I keep alternately kissing: one my new baby brother, the other I'm not sure about. But it is someone I'm glad to see because I am torn between these two who are sitting at opposite corners of the living-room in my house. Someone is crying—I think it's my mother; but (for once) I'm not alarmed by this. I feel elated and the centre of attention. There are other people in the room too and, in spite of this definite, isolated area of sadness (or, more specifically, crying), the general feeling is a happy one. (Age 3.4)

- A cramped, enclosed space, dimly lit. My grandmother, in a dark dress patterned with tiny white flowers, sits at an old green-baize card-table knitting. My two brothers lie stacked one above the other on rough wooden bunks, wrapped in grey blankets. I am in my blue carry-cot. (Age 3.5)
- Being in my bedroom dressed only in a pair of knickers which I had filled with plastic Noddy/Big Ears figures and my mum being really angry with me. (Age 3.5)
- Standing by our back door watching a collie eating from a bowl when it turned on me . . . Blood. General panic. Everybody blaming each other, but making a fuss of me. Apparently I had been standing between the dog and its master, a shepherd. It bit me perilously close to the eye. (Age 3.5)
- Seeing from the butcher's shop, where my mother was making a purchase, a sports car screech to a halt at a zebra crossing but not in time to prevent an elderly woman being knocked to the ground and Bill Drogan (I remember his name) the butcher's assistant running out with bloody hands (from the meat!) to assist the woman. (Age 4)
- I remember my brother being born. I was about four. My brother Mike is four years younger than me—he was born September 1958. What I do remember is him being in a bassinet and his head was shaped in some weird way. 'He's got a weird-looking head.' That's what I remember. (Age 4)

Interestingly, the last memory above is Richard McNally's, whose earliest memory is consistent with his belief that we can't remember much before the age of four.

The first question that might arise on reading this collection is: are there qualities of the memories themselves that tell us how true or accurate they are? Madeline Eacott is a memory researcher at Durham University who devises psychology experiments to try to answer that question. When starting this research, she tried to revisualize some of her own childhood memories:

I've got a file where I've just thought of my own earliest memories and I've made a list of them, only one sentence for each of them to cue me, and tried to estimate the age I was, and I've got a list of seventy. What I discovered flipping through them this morning was that they were all definitely memories when

I wrote them down but because I've read the list several times, I now remember writing it and I'm not really clear that I'm remembering the original event. I think that you can do that—if you constantly rehearse them and think about them a lot, you no longer remember them, you remember remembering them.

When we listen to people's memories, not just from childhood but describing the events of last week or yesterday, the common-sense attitude is to treat them as if the teller was replaying a videotape of the events. It can certainly seem like that, if you are the narrator. You project the pictures and sounds onto an internal screen in your head and describe what you 'see'. As I'll try to show in this book, all modern research into memory reveals that this idea is wrong. It may not matter if we nurse this mistaken idea in our everyday lives (although arguments about who agreed to pick up a Chinese meal on the way home are never pleasant) but, unfortunately, this outdated belief can surface in courts of law with more serious consequences than a marital tiff. One of the goals of psychological research into memory is to search for markers in remembered accounts of events that might help us distinguish truth from falsehood. Without independent corroboration, it is a surprisingly difficult task, although Madeline Eacott was pleased to discover such corroboration, as she told me:

My own earliest memory is of hitting my younger sister, and the funny thing is that after I did this she cried and my parents came and naturally, as a child I hadn't thought through the consequences, and I thought 'Oh dear, I'm going to be in trouble now.' But in fact when my parents came and said 'What's the matter?' she said 'A cow bit me.' And I thought 'Phew, got away with that one.' Years later, my mother told me this story but she didn't know that I'd hit my sister. She told me it as a funny thing my sister said. So I can verify at least part of the story because she didn't know the other part of it. That is one of my memories that is on my list that I definitely remembered at the time, but now I've told the story a couple of times and it's like a funny story I can tell but I no longer have that sense of remembering. It's gone because I've rehearsed it so many times.

Charles Darwin's earliest memory featured a cow too, a real one, and when he wrote it down he also gave reasons why he thought it was an accurate memory rather than one that had been shaped by family reminiscence:

My earliest recollection, the date of which I can approximately tell, and which must have been before I was four years old, was when sitting on Caroline's knee in the dining room, whilst she was cutting an orange for me, a cow ran by the window, which made me jump; so that I received a bad cut of which I bear the scar to this day. Of this scene I recollect the place where I sat & the cause of the fright, but not the cut itself.—& I think my memory is real, & not as often happens in similar cases, from hearing the thing so often repeated, one obtains so vivid an image, that it cannot be separated from memory, because I clearly remember which way the cow ran, which would not probably have been told me.[2]

Darwin, being the observer and thinker he was, had already put his finger on the factors that bedevil attempts to assess the accuracy of very early memories—the role of repetition, usually in the context of family reminiscing, which I will deal with later in the book.

One small aspect of these memories which I'll return to later is the fact that the dating of them is all done by indirect means. Memories don't appear to be date-stamped. Sometimes, people have two or three memories that they can't place in order—they all seem like the earliest. So we have to try to remember which house we lived in at the time, and work out when that was. Or, in the following memory, for example, it was the process of trying to date it that revealed that it could not have been a true memory at all:

• I used to imagine that I could remember my father playing with our labrador dog. It was actually fond imagination based on a photograph. I was only 21 months when he left to fight in Egypt never to return.

As the age of first memory increases, the memories in my sample become richer on average and more sophisticated. Of course, we can never be sure that these aspects of memory narration were not introduced at a later stage:

• My father picking a fruit from a tree in our garden, and telling me its name: 'greengage'. He could reach the fruit easily because he was so tall. Being put to sit on the end of my baby sister's pram, facing towards our nanny, during a walk. Autumn leaves on the ground. Being hit (accidentally) on the head by a sharp crescent-shaped garden tool, with which my brother was playing, when I was standing near him. Having it bathed by my mother with cotton wool and water. (Age 4.25)

- Every morning we would dash down the drive. Halfway along we crept along the strip of grass at the side and carefully pulled away the branches and bracken that camouflaged a large hole. Nothing was ever there. We, my uncle and I, had dug a huge pit—it was our war effort. If the Germans invaded and marched up the drive to capture us, they would fall into our pit and we would be saved. We called it 'the Hitler trap'. (Age 4.5)
- I remember being told I would soon be starting school, but not understanding what this meant. In my puzzlement all I could do was associate the sound of the word 'school' with that of 'nail'. I knew what a nail was. Clearly this didn't get me very far. (4.5)
- Nursery school, 1950s, after lunch getting out mattresses and cots, normally stacked at the side of the room, then lying down for a rest and having stories read to us. The only book—and there must have been many—that I remember very clearly is John Bunyan's *Pilgrim's Progress*—frightening and extremely dark. The image of myself as a child lying down and hearing about Christian's travails has often come back to me in later life in different ways especially the eagerness with which I longed for the next instalment: did he survive? (Age 5)
- One early memory is of my father getting so soaked on his way back from the bus that he had to take all his clothes off in the living-room. He leapt around laughing, dangling all over the place, which my mother and I found very funny and also, in my case, rather rude. But there occurred to me at the time the image of him walking down the path at the side of the house to the bus stop a few hours earlier, when it was such a nice, sunny day that he hadn't bothered to take a mac or umbrella. It seemed a somewhat cruel contrast with the ferocious rain that had come out of nowhere and I saw it as a warning that things can change rapidly. (Age 5)

In the sample of memories in this chapter, there seems to be a fuzzy transition between the very earliest, which are brief, sense-based, and imagistic, and the memories from an older age which tend to have more detail, an episodic quality, a narrative structure. There is also an increasing role played by language in the events of the memory, the thinking about its content or the labelling of different feelings or emotions.

Whether or not these collected memories have any representative value at all as a subset of the adult British public, what we can say is that they are similar to the reports that one reads in other surveys and in autobiographies. If it were possible to have accurate memories of early childhood, the memories I've quoted have the characteristics you would expect—visual and aural glimpses, the involvement of close relatives, usually parents or siblings, feelings of happiness, sadness, or fear, small convincing details. But they also include the unexpected— grandpa's porcelain eye, the knickers full of Noddy and Big-Ears figures. There is a lot of thinking displayed in some of these memories. Even very young children try to make sense of the world with their still developing intelligences and linguistic skills. The two-year-old who thought 'Here I am again' (if he did) is one example, and I came across two other memories, not in my survey, which, if true, show a precocious sense of self-awareness.

The novelist A. S. Byatt in a BBC radio programme described what she believed to be her earliest memory:

My earliest memory is lying on my back, inside what must have been one of those big perambulators with the hood up and bedclothes up to my chin and sort of measuring with my eye the square corner of the edge of the hood against the sky. I was looking out from under the hood—I didn't know any words like 'hood' or 'sky', so the memory is a purely biological one—with no words attached to it—of the kind of sharp edge of the square line of the hood against the blue sky. The verbal conclusion to which I myself without talking to anybody have come is that this was a moment of some step in consciousness when I realized I was I. I think it was the first memorable moment of self-consciousness, and when I say that, the description feels terribly inadequate for the memory. The memory is in my body very, very clear and I think my body recognises this as some basic moment from which I start off.[3]

A. S. Byatt's memory is very similar to that of the psychologist, Fritz Heider:

I must have been about two years old when I had a flash of insight about having a self, about being a person. I was sitting on a little footstool in a sort of nook formed by the doors between two rooms. My father was reading to my older brother, and I felt left out, frustrated. I remember the awareness that it was I who was frustrated, and this discovery of the self made it an exciting experience.[4]

A second sharp sense of awareness, also involving language, was in the earliest memory of Jonathan Schooler, a leading memory researcher at the University of British Columbia. In spite of the fact that many researchers believe that children younger than three or four years can only have very basic visual memories, Schooler's memory had a narrative structure and some quite deep intellectualizing, at least for a two-year-old:

This is a memory that I have been recounting ever since I could talk, my parents tell me—they are both psychologists and this is a memory they would never have told me. I was lying in my crib and I remember crying for my mother, and I remember frustration. I now attribute that frustration to feeling that it was a failure of communication. I just thought she didn't understand that I wanted her to stay there and if I could just get her to understand she would stay there by my crib for the entire night if only I could communicate. She just didn't get it and then I just remember crying and crying and then she left. I just thought it was a failure of communication not that she wasn't interested. What I didn't understand was that she would just go 'No, I'm going back to bed.'

Schooler's first memory seems very much like a small child striving for the concept of language before he even knows what it is.

It's perhaps not surprising that the child of a pair of psychologists should remember in detail an episode from as early as two years old. Or, to put it more accurately, should *believe* he can remember such an episode. We'll see in the next chapter that attempts that have been made to study such memories systematically share McNally's scepticism about how early we can truly remember our childhood experiences. But even from these anecdotal and personal accounts there is much variability.

I conducted an informal experiment with a friend who was visiting me for the weekend. It so happened that this friend had been brought up for the first few years of his life in a seventeenth-century house in a town near where I live. Furthermore, he had not seen the inside of this house since he left it at the age of two. Before visiting the house he told me that his earliest memory was of befriending a Polish prisoner of war who had been a gardener at the house. This memory—if genuine—could therefore be assigned fairly safely to an age of not much older than two. He was interested—and so was I—to see whether the experience of visiting a place that at the time had been so important to him would trigger any memories that he had not had since.

The house today is a tourist attraction and it now receives a regular stream of visitors to see it furnished and decorated in the style of its seventeenth-century occupants. But the location, sizes, and shapes of the rooms would be the same as they had been in my friend's child-hood, and we both expected, and certainly hoped, that there would be an 'aha' experience, as he walked into a room that might have been his own bedroom or a familiar sitting-room, and in a Proustian moment—treading on an irregular floorboard, perhaps—a whole era would come flooding back.

From hallway to dining-room, dining-room to kitchen, kitchen to sitting-room, up creaking stairs unchanged for hundreds of years, to the master bedroom with windows looking out on to the road, and then several other smaller rooms overlooking an ancient mulberry tree in the garden, we walked with the house's custodian. But not a single childhood memory came back to my friend. The most that he could dredge up was a partial belief that he could remember the mulberry tree, which had certainly been there when he was a child as it dated back to the seventeenth century.

One suggestion for why my friend remembered little or nothing of a place that was such an important part of his childhood is that all the things we remember vividly, even apparently trivial things, have some underlying context or significance that is really the important motivation for remembering. We remember the leaves against the sky, for example, not because there was anything significant about the leaves themselves but perhaps because they are a marker for the first time we were put out in the open in our pram and sensed with awe the vastness of the outside world. The psychiatrist, Donald Spence, wrote:

[P]eople often will feel dissatisfied when . . . they go back to their elementary school or their college fraternity—because looking at the real object does not substitute for remembering it, because the remembering, often inadvertent, is embedded in a specific situation with its special context and its own dynamics. A trip to the fraternity house is simply a trip, and it produces a sense of disappointment because it is deprived of the critical surround.[5]

One final sceptical comment on the genuineness of people's accounts of their very early memories comes from a leading memory researcher, Daniel Schacter, at Harvard, who draws attention to the role of suggestion, repetition and imagination in fixing in our minds what we believe are our earliest memories:

There is no evidence that people can remember incidents that occurred before they were two years old, most likely because the brain regions necessary for episodic memory are not yet fully mature until that age. In one recent study, people generally reported earliest memories from when they were three or four years old, as in most previous research. The experimenters then introduced a suggestive procedure in which they asked subjects to visualize themselves as toddlers and try to 'get in touch' with even earlier memories. They offered assurances that just about anyone can remember very early events, such as a second birthday, by 'letting go' and working hard to visualize the event. Following the suggestive procedure, people reported earliest memories that dated, on average, to approximately eighteen months—well before the accepted offset of childhood amnesia. Indeed, one-third of those exposed to the suggestive procedure reported an earliest recollection from prior to twelve months, whereas nobody did so without suggestions. Because there is no other evidence that people can recall events from this early in their lives, these newly discovered 'memories' almost certainly do not reflect accurate recall of events.[6]

When Schacter says there is no evidence, he is talking about a very specific sort of evidence, that produced by well-organized, rigidly controlled scientific experiments. So what do the experiments that *have* been done tell us about the nature of childhood memory and childhood amnesia?

2

· · · · · · · ·

Childhood Amnesia

Here are two statements about childhood memory. The first is by a psychotherapist:

Some clients remember back to six months old, when their fathers may have taken them out of their cribs and molested them. One of my clients remembers being dropped on the floor after being molested and dislocating her hip. That memory came back to her at home, in a meditation she did on her own.[7]

The second is by a research psychologist:

An event that occurs when a participant is younger than 3 years old may be recalled in adulthood. Moreover, many participants are able to show substantial recall of events that took place when they were age $2\frac{1}{2}$, but recall of events that took place in the first quarter of this year are much more rare. However, in situations where memories from this earliest period are recalled, we have no evidence that they are less accurate than memories from the later period. Thus, we are able to point to a steep offset of childhood amnesia during the first half of the 3rd year of life.[8]

The first statement says that people can remember events dating back to when they were six months old, sometimes as a result of meditation as an adult; the second that it is unlikely that anyone can remember any event that took place before the age of two and a half.

These statements fundamentally disagree. Maybe this is not surprising. Scientists are always arguing among themselves, aren't they? But there are clues in each of them that suggest that one is much more credible than the other. The first is brief, unqualified, anecdotal, uncorroborated, and arrived at by a decidedly peculiar method. The second, as we will see, is cautious and more detailed and based on the result of months of meticulous experimentation, statistical analysis, and careful wording.

The problem with studying memory is that we are all capable of making inferences about it, since we all use our memories. It is all the more important, therefore, that when statements are made, they are backed up with evidence rather than the kind of untested personal impressions that feature in the psychotherapist's statement above.

Memory is one of those subjects where, as a researcher, it's difficult to be objective. As a human being you have views about your own memories which could get in the way of accepting conclusions that disagree with them. You must therefore be a professional sceptic. A psychotherapist, however, would lose clients if she went through life disbelieving the stories brought to her by distressed people seeking her help.

Elizabeth Loftus, a leading memory researcher, takes a sceptical attitude to almost every early memory that is offered as genuine, even—as we've seen—her own. In 1993 a key research project in the field, by JoNell Adair Usher and Ulric Neisser, tried to pin down the duration of childhood amnesia by finding the earliest age at which people could reliably be said to have a memory for life events. The study claimed that people could remember a stay in hospital or the birth of a younger sibling that occurred when they were as young as two or two and a half. Loftus views that study with a degree of scepticism, as she told me:

They based this on the fact that people who had a sibling that was born when they were two plus could answer some questions about it, and the questions were things like 'Who took your mother to the hospital?'—'Daddy', 'Who waited with you while your mother went to the hospital?'—'Grandma', and I'm sitting here saying '*I* can answer their questions.' This was not very good evidence that really these people were remembering these things.

In science, when the results of research are interesting but not watertight, the way is open for other scientists to repeat the experiment in a way that tightens it up and removes any reason for scepticism.

Madeline Eacott did exactly that with the Usher/Neisser research. She is a psychologist at Durham University who started her research career in the rarefied and specialized science of neurophysiology and then broadened to behavioural work. She described to me how the starting point for her research was a desire to test a common-sense idea that most people believe:

When I had sometimes written to parents saying 'Can you tell me whether your child's memory is correct?' I sometimes get back people saying 'Well, obviously she can't remember, she was only two.' And they've taken it entirely for granted that of course you can't remember. So there's this general accept-ance out there 'Of course, you can't remember.' And you think 'Yes, but why not?' Nobody seemed to question that. If you ask a hundred people they'll say 'It's obvious that you can't' because they'll say 'It's our experience that you can't', even though most people aren't necessarily familiar with the term infantile amnesia or childhood amnesia.

In an early research project, Madeline Eacott decided that when her two children were very young she would do what researchers call a longitudinal study—collecting data about the same people over a long period to look for changes and correlations that might not be observable in the short term. (Psychologists have a history of using the subjects who are closest at hand, their own children.) She was interested in particular in the difference between two types of memory—semantic and episodic. Episodic memory is what makes up the bulk of people's early memories—specific places, people, and events. Semantic memory concerns memory for words and concepts and general knowledge about the world. The content of early memories is not usually explicitly semantic—apart, perhaps, from the boy in my survey who wrestled with the idea of the word 'school' and tried to relate it to the more familiar but unconnected word 'nail'.

Eacott wanted to see whether semantic memory was also subject to childhood amnesia, or whether very young children had semantic memories that related to an even earlier period in their life than two or two and a half.

I thought, perhaps my children do remember all sorts of stuff, it's just not explicitly available. So we deliberately exposed them to various materials when they were very young for a period of time, six months, then simply removed the materials from their experience and tested them again when they were aged eight or more to see what they remembered. Had it gone, like infantile amnesia, or was it still there either explicitly or implicitly?

To make sure that what the children had to remember was not something they might come across later, before they were retested at the age of eight, Eacott chose abstract patterns to which she gave nonsense word names. One shape might be a 'rorp' for example, and the child would have to learn that connection:

They were exposed to twenty patterns and their names which they learnt, and we played games with them, we played snap with them, we played dominoes...I did it from age two, within the period of infantile amnesia, two to two and a half, and they were highly familiar with this stuff for a period of six months and then we removed it, and we had a control group exactly the same but they hadn't been exposed. We tested their explicit memory for it, but there was also another set of tasks where they'd be shown a pattern and I'd say 'what would be a better name for that?—would it be *rorp* or something else.' We'd say 'We know you don't know but just guess what would be a good name for that,' and we wondered if they would pick the name that they'd learnt as children.... What we found was that despite the fact that they had been highly familiar with this stuff as children over a period of six months, when presented with it later, there was nothing, certainly nothing explicit. There is the slightest hint of something there, but it's not very much, so it's really very difficult to really be clear what the answer is...

A scientific experiment doesn't always produce the result you hope for, but if it's well designed, it should produce new knowledge. The fact that Eacott's children had no dramatic recall of the shapes and their names suggested that both semantic memory and episodic memory had a period of childhood amnesia.

But what about the Usher/Neisser results and Loftus's objections to them? Could Eacott design an experiment to discover whether such earliest memories are actually the result of guesswork or family knowledge rather than true memory?

Madeline Eacott and her colleague Ros Crawley decided to repeat the Usher/Neisser work, but with an additional element that ruled out alternative explanations for the results. In the earlier experiment, students were asked about memories of events associated with the birth of a sibling at some time between the ages of two and four. Questions like: Do you remember being told that your mother was going to have a baby? Who told you? Where were you? What were you doing when she left? Who took care of you while your mother was in hospital? How did you find out that the baby was a boy or a girl?

They were told to exclude memories which might have been triggered by conversations with relatives or looking at family photos, but there was no independent way to verify whether or not, consciously or unconsciously, they were basing their memories on that sort of information. When the students' memories had been gathered together, each account was sent to the student's mother who was asked if she could corroborate it.

Eacott and Crawley hoped to rule out the influence of family reminiscing with an ingenious experimental design.

Their subjects were drawn from among the psychology students at their university. (It is sometimes remarked that everything researchers have discovered about psychology really only applies to rats and psychology students.) One group of students, called the 'recall' group, was chosen from those who had siblings who were two or three years younger. These were then asked the same questions about events surrounding the birth of the sibling as those asked in the Usher/Neisser study, and their mothers were asked for corroboration. As with the earlier study, the researchers found the same age of onset for childhood memories (or offset for childhood amnesia, which is the same thing). But, to act as a control group, called the 'report' group, Eacott and Crawley took students who had siblings who were *older* than themselves, and asked them similar questions about their own births. So where a member of the Recall group was asked who looked after him or her at the time of the sibling's birth, a member of the Report group was asked who looked after the older sibling when he or she was being born. Any information the Report subjects had about their own births—who looked after their sibling while their mother was giving birth, what presents their sibling might have been given and so on—could only have come from outside sources rather than their own memories for those events.

The results of the Eacott/Crawley experiment showed that the two groups—those who were relying on family reports of their own birth and those who had access to those reports as well as to their own recall of their sibling's birth—could do equally well at telling stories about significant events in their childhood, whether or not they had actually remembered them. So the corroboration from a mother that a story was correct didn't necessarily mean that it was a genuine memory.

But when Eacott and Crawley looked more closely at the two types of memory, they found very significant differences. The 'reported memories', connected with a subject's own birth, had more detail. This was a surprising result on the face of it. You might have expected the Recall group, remembering the birth of a sibling, to produce more details since they would have access *both* to their own memories *and* to family stories or photographs. But, for Eacott and Crawley, this less detailed nature of the recalled stories was an indication of their genuineness: 'It is unlikely,' they wrote, 'that our recall participants were simply reporting to us all the information they knew, being unable to discriminate between sources of knowledge.' They went on:

If that were the case, our recall participants would be expected to relate the same amount of information as our report participants or perhaps even more, as they could supplement family knowledge with true autobiographical memories. However, this was not the case. Thus the recall participants filtered their knowledge about the events and reported only a subset of this knowledge.... [T]he conclusion from this discussion is that although participants have a great deal of information available to them about their sibling's birth from a variety of sources, they are not reporting all this information as though it were autobiographical memory.[9]

I have gone into this research in some detail because it is the first of a number of key experiments I will describe in this book to show how carefully experiments have to be designed to allow us to make firm statements about human psychology. We live in a world where we are all our own experimental subjects and can all have theories about what makes us tick without always taking the care—or having the ability— to separate out fact from myth, hunch, or wishful thinking. With research like the Eacott/Crawley experiment, we can accept with some degree of confidence that childhood memories back to the age of two are possible if sparse. Also, they say at the end of their paper:

Perhaps the most important aspect of these data is the finding that those who were aged 2:4–2:7* and older at the time of their sibling's birth have significantly more memories of the event than those who were aged 2:0–2:3.[10]

In a rough and ready way, this effect is borne out by my informal survey, where the memories claimed at between two and three are shorter and have less detail and less narrative structure than those after the age of three.

Up to now I have been using the word 'memory' in a very basic, non-scientific, sense. For scientists, there are many different sorts of memory. Even in the first couple of years of life the child is learning and retaining a huge amount of 'knowledge'—he or she is 'remembering' how to eat, how to sit on a potty, how to direct the gaze towards a sound, how to recognize a specific human face, how to crawl, and so on. Clearly, we have these 'early memories' for the rest of our lives. Humans also have good memories *during* their early childhood. The fact that even the memories we have from age three or four or five are a tiny subset of the events that happened doesn't mean we don't have good memories at the time. From day to day, babies and toddlers store up a lot of information about events and people which they can recall a few days later. But, as we've seen with Eacott's experiments on her own children, these don't persist beyond the immediate short term.

If we accept that there is a period of childhood amnesia up to two or two and a half, but that there are some genuine memories retained until adulthood from the period from two and a half onward, there are two questions it is interesting to ask:

* The use of '2:4' to mean 2 years and 4 months old will occur from time to time in this book.

Why is there an early period which is not remembered at all?

What is it about the small number of memories that we *can* recall later from early childhood that makes them memorable, when so much of our early lives is forgotten?

The inventor of psychoanalysis, Sigmund Freud, made one of the earliest attempts at an explanation for childhood amnesia. Not surprisingly, he thought it had something to do with sex, and that the earliest memories were blocked by a wall of repression. He wrote 'It is impossible to avoid a suspicion that the beginnings of sexual life which are included in that period have provided the motive for its being forgotten—that this forgetting, in fact, is an outcome of repression'.[11] But suspicion is not evidence, and there are no reliable data supporting Freud's 'explanation'.

There are two main hypotheses offered by modern psychologists for why we don't remember our first two years. One is that the brain is still developing and the proper functions of the nervous system for acquiring and retaining memories are not in place; the other is that we are very limited in what we can store and recall until the age at which we develop language.

The two explanations are not independent. Language and brain development go hand in hand, so if memory was dependent on language ability which in turn depended on brain maturity, then brain development would be a factor. On the other hand, if memory ability does not depend on verbal ability but on some other aspect of brain function, then we could have pre-verbal memories if the brain systems were in place to process them.

There are many books and papers which describe theories of memory based on brain anatomy and physiology, but apart from certain very basic statements about specific areas of the brain 'playing a role' in memory, there is no generally agreed and detailed explanation in terms of brain organization of what happens when an outside event is perceived, stored—for a short or long period—and recalled. Much of the information about how the brain works has come in the past from testing patients who have suffered damage to different parts of the brain. But such research doesn't always give precise information about the mechanisms involved. To take a trivial example, if the on–off knob on a television gets broken there is no longer any picture or sound, and we might think that the essence of the workings of the television reside in that one component, when, in fact, the inside circuitry might still be functioning perfectly. (In recent years devices which can measure the activity in different areas of the brain, such as functional MRI detectors, have begun to open up the real-time functioning of the brain to outside analysis but there is still much we don't know.)

Starting with the input to our memory, we take in information through all five senses, and this information travels along many different pathways to parts of the brain, activating different areas depending on the importance and nature of the event we are perceiving. It can be quite difficult to interpret why someone's memory has gone wrong on the basis of random injury to the central nervous system. Nevertheless, this kind of research is used to draw diagrams showing areas called the pre-frontal cortex, labelled 'working memory', the hippocampus, 'long-term memory', the cerebellum, 'skill memory', the amygdala, which is involved when memories have an emotional content, and so on. And some kind of consensus seems to be emerging about the importance of one specific area of the brain, as described in a recent paper in *Science*:

The frontal regions are critically involved in manipulating and organizing to-be-remembered information and in devising strategies and monitoring for retrieval, although the frontal regions themselves may not be intimately involved in the binding of information into long-term memory.[12]

The frontal regions of the brain, sometimes called the frontal lobes, take up the part of the head behind the forehead and beneath the skull. They seem to act partly like the Issues Desk at the entrance to a library, dealing with loans and returns, cataloguing new books as they are acquired, deciding where to shelve them, and dealing with readers' enquiries about where to find specific books in the library. To give an idea of how unspecific our understanding of functional brain anatomy still is, a recent encyclopaedia entry says that 'the frontal lobes have been found to play a part in impulse control, judgment, language production, working memory, motor function, problem solving, sexual behavior, socialization, and spontaneity'.[13] Clearly, these are powerful—if rather unspecific—areas of the brain, and it's easy to imagine, if not to prove, that until these lobes are fully developed in the child there's an awful lot we can't do, including store memories.

Madeline Eacott, who considers herself at the biological end of psychology, says, with a laugh, 'I tend to regard all of childhood as a disorder of frontal lobe function', but even she can't quite put her finger on the brain mechanisms that underlie memory:

I think the ultimate answer to infantile amnesia will not be 'And the winner is . . .' Development of the frontal lobe would only allow cognitive processes to happen which couldn't previously happen, so you could say the cause of it is lack of frontal lobe development or you could say it's the lack of ability to code personal memory, but they could both be correct, just different ways of looking at the same thing—it's like the story about the elephant and whichever bit you happen to be feeling, it's the same elephant. . .

There are several explanations for childhood amnesia which address different stages in the memory process. It could be because a key centre in the brain, the hippocampus, which plays a part in the *recall* of memories, is still immature. Or, connections between the areas of the brain that store memories for a short time are not yet connected to the long-term storage area. Or, very early memories don't have much contextual detail of the sort that enables us to form memories later in life, the 'hooks' that connect our memories to different parts of the brain, so that we remember a particular painting because it belonged to a good friend, it showed a picture of our favourite city, it cost an awful lot of money, or it fell on our head one day. Or—another possible explanation—memories that were formed before we had words to describe them cannot be retrieved at a later stage when we have acquired speech, almost as if having read something in French at an early stage, we couldn't retrieve it if we later replaced French with English.

To return to the library analogy, there could be several reasons why a particular book does not appear to be available:

It may never have been acquired by the library.

It may have been acquired but not catalogued.

It may have been catalogued but mis-shelved and so irretrievable.

The Issues Desk may be closed.

The book may have been acquired once but disposed of to make space for more recent books.

It is only available in a foreign language.

All of these would justify a potential borrower going away empty-handed, just as the defects of memory based on the roles of different parts of the brain could all have the same effect, the unavailability of memory for a particular event.

In the end, I find descriptions of behaviour in terms of anatomy and physiology rather unsatisfying. It's like explanations sometimes offered for the mysterious phenomenon of sleep—'It's to allow the nervous system to rest', they say, but then I want to know: why does the nervous system *need* to rest? Could evolution only shape a system that functioned for sixteen hours out of twenty-four? This 'explanation' just seems to replace one mystery with others.

We can gain more useful information about memory from research into cognition—how ordinary people think and speak and behave—than from looking at the anatomy and physiology of the brain. Susan Clancy, a research psychologist at Harvard, explains:

The critically important point is this: our brains engage in an act of construction when we call up our memories. There's no single place in the brain that contains a 'memory chip' of an experience. Instead, when we try to recall an experience, a reconstruction process begins. Various parts of the brain that have held on to disparate aspects of the memory work together, and what emerges from this reconstruction—your memory—is not the same as what actually happened. A memory is not an exact photograph of an event. It is created out of the cues that elicited the memory and the fragments of the original experience that were stored in the first place.

This is a creepy idea, and many people understandably resist it. Our memories constitute who we are. They inform our personal history, our life stories, our sense of ourselves. Our lives, after all, are only what we remember of them.

It's unnerving to realize that our stories, feelings, memories of the past are reconstructed over time, and that we make up history as we go along.[14]

Looking at the memory in terms of behaviour and cognition, rather than anatomy and physiology, can help explore the suggestion that childhood amnesia is connected with the development of language and verbal behaviour.

One view of language and memory might suggest that it is possible for a small child to preserve a memory of a scene, say, or an event, that was experienced *before* language developed and then 'replay' it and describe it with words she later acquired.

On the other hand, as described by David Pillemer, a researcher into autobiographical memory, there may be a fundamental barrier that separates the two types of childhood memory:

Events that are experienced before about age two are encoded imagistically. When the child begins to engage in conversations about the past, events are also encoded narratively; but if the young child's understanding of the episodes is limited, and if the child has not yet internalized mature models of narrative structure, the verbal memories may not resemble an adult's conception of what transpired. Deliberate attempts to recall the event years later may fail because the imagistic or childlike memory representation is incompatible with the adult's purposeful reconstructive efforts.[15]

On this theory, for example, language would help in some way to 'fix' a memory in the brain, so that if you can describe something in words you can remember it better than just storing separate images and sounds, while the pre-verbal child is just a conduit through which the outside world passes without any real 'episodes' being captured or stored. This would explain why many of the earliest memories are more like silent snapshots than film clips with sound. Language might also play a part as the developing child discusses the present and the past with his or her parents. The words they use in conversation might help to preserve memories that would otherwise be lost or fragmentary.

There are some interesting experiments that cast light on the role of language in childhood amnesia and confirm the idea that it could play an important part.

Gail Goodman, at the University of California at Davis, has studied childhood memory as part of an interest in child sexual abuse and the effects it can have on children. As some examples in Chapter 1 showed, understanding childhood memory and how it works is particularly important when those memories involve crimes or abuses which might have long-term consequences. But investigating this aspect of memory is not easy. Doing psychological experiments on children is fraught with ethical problems anyway, but studying anything that is related to trauma or sexual abuse, particularly if you want to control the conditions of an experiment, is even more troubling.

Goodman and her colleagues carried out one experiment which looked at the memorability of a traumatic event that took place in children of different ages:

We were looking at a stressful medical procedure that involved urethral catheterisation and we found kids who'd had that in their earlier childhood and then we interviewed them later. They ran all the way up to thirteen years of age, and basically we found that if the procedure occurred when they were two years of age there was either no memory at all for it or there were just a few kids who'd had these fragmentary image-like memories, where you couldn't even be positive they were talking about that medical procedure but it seemed as if they probably were. Then, when they got to be around three years of age when they had it, it's clear they can talk about it and remember it, and as they get to be four and five when it occurred you start to see less of these vague image-like memories—they either remember it or they don't.

In a similar project, psychologists interviewed children at a hospital after they had received emergency treatment for injuries. The children were between one and three years old, and the researchers went back to them at approximately six-month intervals afterwards to see what they could remember.

'Children who were under age two during the initial interview,' David Pillemer wrote, summarizing this work, 'usually could not give a verbal memory, and they did not provide coherent narrative accounts during later interviews. In contrast, slightly older children who were able to describe their traumas in words during the first interview often could recount major aspects of their ordeals two years later.'[16]

So the idea of preserving the pre-verbal memory and then 'playing it back' and describing it in a vocabulary you have acquired later is not really borne out by these experiments.

Recently, the matter has been settled conclusively—until someone comes along with a different set of results—by Harlene Hayne, an American researcher working in New Zealand. She has done a series of experiments with children to try to discover whether they can describe events that took place at time A, when questioned about it at a later time B, using the language skills they had acquired in the interval between the two dates. Her hypothesis was that memories can only be described by children using the level of language they had at the time they had the experience, with the corollary, of course, that if they had no language skills at all they couldn't describe, or even remember, the event.

The experiment called for very young children to experience a unique event, one they could not encounter again as they grew older and their language skills improved. So Hayne invented an ingenious toy called 'the Magic Shrinking Machine'. This was a large and impressive box, with a handle that made interesting noises and lights that flashed. When the experimenter put a regular-sized toy in a hatch on top and operated the machine with the lever, the child was shown how to retrieve a much smaller but identical toy from the front of the machine. This must have been a truly memorable—and certainly unique—event.

The procedure was carried out seven times on two successive days, 'magically shrinking' a range of different toys, and the children were encouraged to operate the machine themselves once they got the hang of it.

At the beginning of the experiment and in a succession of visits to the children, the experimenters gave the children tests which measured the level of their language skills and, in particular, tested their knowledge of particular words that could be used to describe the events of the Magic Shrinking Machine experience.

When the children were visited again after intervals of six months, they were tested both verbally and visually about the Magic Shrinking Machine. They were asked to describe what they remembered of the events; they were shown pictures of the machine and the toys to see if they recognized them; and they were tested on the behaviours associated with the machine, such as pulling the lever or retrieving the shrunken toy.

Hayne and her colleagues wanted to discover whether the children would use words that they had learnt *after* the 'unique experience' when they described their memories. The pictures and the behavioural test didn't depend on verbal ability, so those results would confirm how many children had remembered the events and how accurately. If verbal descriptions of the event were as sophisticated as the children's level of speech when they were tested, it would suggest that they could 'play back' the memory and describe it as if they were experiencing it again. But, in fact, this was not the case. As Hayne and her colleague wrote:

The results of the present experiment yielded no evidence whatsoever that children could translate preverbal (i.e., nonverbal) attributes of their memory representations into language. In no instance during the test did a child use a word or words to describe the event that had not been part of his or her productive vocabulary at the time of the event. In short, children's verbal reports of the event were frozen in time, reflecting their verbal skill at the time of encoding, rather than at the time of the test.[17]

So what does that mean for the early memories many of us claim to have? The results from Hayne and others suggest that things we think we remember from before we learnt to speak are more likely to be genuine and unmediated if they are visual, impressionistic, and rudimentary. The more they seem like verbally sophisticated stories, the more likely they are to be reconstructions from stories we were told by others, photographs from family albums, or even wishful thinking. (There's one small caveat here—even the most basic visual memory needs words to describe it, so there must be *some* use of verbal skills learned later to describe pre-verbal events.)

Now, some of us may resist this dismissal of our proudly displayed early memories. I, for example, can *see* my golliwog in my mind's eye down the hole beneath the grating (all right, I confess, that was *my* first memory) even though at the time I might not have known the word 'grating'.

But then, another paper by Hayne and a colleague says:

We propose that, unless they are 'updated,' memories for discrete events that occur very early in life are not translated across stages of development. Rather, these memories, if retained, are maintained in the same representational format in which they were originally encoded.[18]

So some very early memories that we express verbally might be genuine but have undergone some kind of updating, where a child tells a parent and at some later stage the parent reminds the child, in words which are now more sophisticated than he might have used at the time, and *that's* what he now remembers.

With reference to my 'golliwog' memory, because I was evacuated during the Second World War my mother was not with me at the time I had (or believe I had) the experience, so the version I now remember could have been as a result of telling my mother, who reminded me of it at a later stage.

On this basis, we can perhaps look back at the earliest memories quoted in the first chapter and see how some are probably accurate, some are probably imaginative and inaccurate reconstructions, and some are a mixture of both. But the work of Eacott and Hayne and others does seem to have pushed back the age at which genuine memories can be recalled. Ten years ago, psychologists were saying that nothing can be remembered before about three and a half. Now, apart from by a few diehards, the idea of accurate memories at two and a half or two is becoming accepted.

But there's another interesting point here. If we do have early memories, they are clearly a tiny selection from the things that happen to us as children, day after day, week after week. Why do we remember the things we do remember? What is special about some of these seemingly trivial events?

This is not as susceptible to scientific experimentation as the issue of whether early memories are accurate. How could we possibly discover fifty or sixty years after the event why someone still remembers 'Waking up in cot, lying, seeing yellow curtains waving in breeze, somebody nearby having a shave'? Or 'Holding a woman's hand entering an apartment, walking in the narrow entryway and seeing a bathtub on the left'.

The psychoanalyst Alfred Adler believed that earliest memories 'have great psychological significance, providing important insights into the very core of an individual's personality'.[19] As we'll see, a therapist's belief about the significance of an early memory can lead both therapist and client down a very dubious path. Some results from research into reading and learning suggests that what we remember, at any stage in life, is not always what is important.

One experiment presented children and adults with two versions of a text about insects. Here's how one version began:

Some insects live alone, and some live in large families. Wasps that live alone are called solitary wasps. A Mud Dauber Wasp is a solitary wasp. Click Beetles live alone. When a click beetle is on its back, it flips itself into the air and lands right side up while it makes a clicking noise. Ants live in large families. There are many kinds of ants. Some ants live in trees. Black Ants live in the ground.[20]

Readers were asked to identify the 'really important information in the text'. Many readers of this passage identified the fact about the click beetle and how it rights itself. But, in fact, it's far more important for the learner to know that some insects live alone and others in large families. When the irrelevant fact about the click beetle was omitted from the passage and replaced with neutral material, far more readers identified the important information. The researchers called this the 'seductive detail' effect, and it's quite possible that this operates with whatever we perceive, not just reading texts, and that many of our early memories are *not* particularly significant, just interesting.

David Pillemer is a psychologist who has written a book called *Momentous Events, Vivid Memories*, with the subtitle 'How unforgettable moments help us understand the meaning of our lives', in which he suggests various reasons why memories of particular events stay with us while others quickly dissipate.

It is possible that some trivial moments stay with us because they were accompanied by a sudden insight that we find useful. The boy in Chapter 1 who saw his father go out without an umbrella on a sunny day and return soaked to the skin, learnt the lesson that 'things can change rapidly'. This also fits with two of my childhood memories. Once, I was discussing with my mother the name of a department store in the high street, which I believed was called 'Spratt's'. I passed this store often on my way to school on the bus and so I was sure I was right. My mother told me that it was actually 'Pratt's', and I just did not believe her—until the next time I passed the store and realized to my chagrin that she was right and I was wrong.

On another occasion, I was about nine and my grandmother was going to take my temperature because I felt off-colour. I offered to take the thermometer out of its metal case and she told me to be careful in case I dropped it. I seem to remember that I was confident that I could do it, and started to pull the top off. It was rather stiff and then suddenly it came off, sending the thermometer flying onto the tiled hearth and smashing into smithereens.

I think it is possible that both of these incidents have led me to try not to be too confident about anything in life unless I have a basis of evidence or experience. 'Never assume' is what my children remind me I drummed into them when they were young. (Conversely, given evidence for a belief, I can be very confident.) The incidents were trivial but memorable; the significance, for my approach to life, was profound.

David Pillemer developed this idea of memories as indicators of more profound aspects of our lives with a story of his own:

Several years ago, my then toddler daughter's beloved blanket was in the wash when I dropped her off at a baby-sitter's house. I reassured her that when I came to pick her up the blanket would be in the car. Later, when I was tempted not to bother with the blanket, I remembered a time when a senior colleague had asked me to order a book for him and, not taking the request seriously, I had neglected to do so. The lesson? Bring the blanket. Yet the concrete circumstances, including the people and types of objects involved, were quite dissimilar.[21]

The ethologist Patrick Bateson, in summarizing the current state of knowledge about childhood memory, describes some interesting work on the transient but useful role of memory in very early life:

A big change occurs between the ages of approximately two and four years after birth, with the emergence of language and an awareness of self. Few adults remember much of what happened to them in their first few years. Even if they are subjectively certain that they remember their birth, say, or some early incestuous molestation, the corroboration is invariably suspect or missing. It could be argued that in the first few years, children have no memories; nothing has been stored so nothing has to be erased. Such a view is clearly false. Young children have good functional long-term memories. In one experiment, for example, children around two years of age were asked to imitate actions they had seen eight months before. They performed significantly better than children who had not previously seen these actions.[22] The absence of memories from infancy does not reflect an inability to form enduring memories at the time, suggesting instead that substantial reorganization of memory occurs early in life.[23]

It's that word 'reorganization' that holds the clue to the purpose of childhood memory and to why we remember some things and not others in a seemingly unconnected way. The psychologist Katherine Nelson, at the City University of New York, has developed a theory of childhood memory that explains childhood amnesia and also suggests a way in which it plays a part in the development of autobiographical memory, the small selection from the many impressions a child receives which collectively give her a sense of self.

Someone who played an important part in the development of that theory—in a very unusual way—is Emily Oster, today an assistant professor of economics at the University of Chicago. A recent paper she wrote deals with a topic that is a long way from childhood memory. It begins:

'I generate new data on HIV incidence and prevalence in Africa based on inference from mortality rates. I use these data to relate economic activity (specifically, exports) to new HIV infections in Africa....'

But among Oster's contributions to the study of childhood memory is something she said twenty-five years earlier. It began:

> 'Car broke,
> The—Emmy can't go in the car.
> Go in green car.
> No. Emmy go in the car
> Broken. Broken. Their car broken,
> so Mommy Daddy go in their their car,
> Emmy Daddy go in the car,
> Emmy Daddy Mommy go in the car, broke,
> Da...da car...
> their, their, car broken.
> So go in the green car.
> Fix the car broken.
> So go in the green car...
> Car fix their car broke,
> So their car broke...'[24]

From the age of twenty-one months, Emily Oster's pre-sleep monologues in her crib were recorded as part of a unique research project into early speech and memory. The research, led by Professor Nelson, was based on 122 tapes of Emily's crib talk, which were transcribed and annotated and then used by a group of researchers to look at many different aspects of early language.

Emily was a bright child and a precocious speaker, but the content of her bedtime musings has cast light on the functions of memory that can be generalized to explain what we all remember or forget of our childhood and how we develop a sense of self.

3

•••••••

How Do I Know Who I Am?

Before baby Emily Oster went to sleep at night, she generally took part in two types of activity. Often, she would have a conversation with her father or mother. This was usually about the events of the day, as one of her parents described what the family had done together, but sometimes the conversation would look ahead to the events of the following day. Emily's contributions to these conversations were often brief, acknowledging that she understood and repeating a particular piece of information. One typical dialogue included the following:

> 'I came down to get the laundry!
> Tell it to me again.
> And everybody went home, everybody's home.
> I goin' to the park.'

After such a conversation, Emily was left on her own to settle down, and it was at this point that she often indulged in long monologues. Now the type of brief statement that she made in the dialogue was replaced by long, more complex musings in which she often spoke about the future as well as the past. This is how a typical monologue began, when Emily was twenty-eight months old:

We are gonna ... at the ocean. Ocean is a little far away. Baw, baw-buh-buh, baw-buh-buh ... far away ... I think it's ... couple *blocks* ... away. Maybe it's down, downtown, and across the ocean, and down the river, and maybe it's in, the hot dogs would be in a fridge, and the fridge would be in the water over by a shore, and then we could go in, and get a hot dog and bring it out to the river, and then sharks go in the river, and [bite] me, in the ocean, we go in to the ocean, and ocean be over by, I think a couple of blocks away. But we could be, and we could find any hot dogs, urn, the hot dogs gonna be for the beach. Then the bridge is gonna, we'll *have* to go in the green car, cause that's where the car seats are. Urn, I can be in the red car, but, see, I be in the green car. But you know who's going to be in the green car—both children.... I'm going to be in the green car in my car seat, he's gonna be ... and nobody's gonna be, just ... you know, th-e-se people, we don't know, and too far way from the beach, and two things.[25]

There were several things that Nelson and her colleagues noticed about Emily's crib talk. First, there was a big difference between her contributions to the conversations with her parents and the content of her monologues. It seemed that the conversation had some social purpose—shared reminiscences that helped the process of bonding with her parents. But the monologues were much more to do with daily routines and any variations that Emily noticed and, perhaps, puzzled over. There were few references to truly novel events of the sort that might stand out in adult memory, and when they did appear, they were not retained for very long and seemed to be forgotten a few months later.

The daily events that formed part of a regular pattern *were* remembered, 'in an elaborate and extensive verbal form', Nelson wrote, and she went on to suggest that the purpose of this process was related to the fact that the young child has a lot to learn about the world in a very short time. Because much of the world is predictable—such things as regular mealtimes, the constancy of parental appearance, the fact that things fall to the ground if you let them go, the regularity with which grandpa pats you on the head and produces a coin from your ear—it makes sense to learn these regularities about the world, so that the next time they start to happen you know how things are likely to unfold and can act accordingly.

These predictabilities of the world and how we behave in it have been called 'schemas' or 'scripts', and a psychologist who has developed this idea in some detail, Roger Schank, is fond of using the example of restaurant behaviour. (This is partly because—as I know to my cost—he very much enjoys being taken out to extremely good restaurants to discuss his ideas.) After people have visited restaurants a few times they begin to get the hang of what's going on. There's likely to be a series of events that we are expected to follow in a certain order—make sure you take money or a credit card with you because payment will be required, don't ask for the bill before you've had your meal, don't bring cutlery with you because it will be on the table, don't eat a big meal just before you go out, and, whatever you do, don't argue with the waiter while he still has an opportunity to spit in your food.

'The general function of memory is to predict and prepare for future encounters, actions, and experiences,' says Katherine Nelson. 'That is, memory as such has no value in and of itself, but takes on value only as it contributes to the individual's ability to behave adaptively.'[26]

It could be that children prepare for future encounters by learning scripts. These may be simpler than adults' scripts but they are still an essential part of growing up. What was happening for Emily, Nelson believes, was a systematic recital, in her mind and out loud, of the events of the day and how they reproduced, or departed from, similar events on other days. And those departures are important. You need to know the varieties of experience that are still within the bounds of normality. It would be embarrassing if you went into a restaurant and complained about being given two sticks to eat with instead of a knife and fork. But if you had learnt the script, which says that in a restaurant called 'Golden Dragon' or 'Mr Wu' or 'The Rice Bowl' you can expect to be given a pair of sticks to eat with, you will be less likely to be embarrassed.

'Under this view,' Nelson writes, 'the formation of general schemas of routine events is of highest adaptive priority. Noting the variations in routines is important to the organization of schemas that will provide reliable expectations about future encounters. Highly novel events—ones that do not fit prior schemas—may be retained, available for schematizing when another such episode is experienced. However, if another episode similar to the first does not appear, memories for these events will have no apparent adaptive value. We have seen that in fact novel episodes play little role in Emily's memory talk. I have also suggested that she talks primarily about events that need organization and resolution. When a single novel episode occurs, there is no way to organize and resolve its similarities and differences with other events.'[27]

It was even possible to test this theory by doing an analysis of the content of Emily's utterances between the ages of twenty-one and thirty-six months, and seeing how it varied from month to month. At the beginning of the period, Emily talked about sleeping, eating, going to the baby-sitters, and her grandmother coming. But these topics declined over the next six months.

'Presumably,' writes Nelson, 'talking about these topics in the monologues enables her to get them under control, in the sense that she understands the content and structure of these daily events and is able to predict their occurrence and sequence. When this level of understanding is achieved, these topics are abandoned, and the monologues then focus on different types of material.' (Note that the other topics didn't drop out because they were no longer happening—she was still eating, sleeping, and so on.)

But then, at about thirty-one months, the exploration of routine events in her monologues increases, as Emily experiences the new world of nursery school. This time, she 'organizes' the material more quickly, perhaps because she's 'learnt to learn', as psychologists say.

And against this background there is really no place for novel memories. At least not for memories that remain outside any conceivable schema or script. There will be novel events, of course—the child will taste avocado for the first time, or sleep in a new crib—but those novel events will soon be assimilated into an 'eating' or 'sleeping' script. The *truly* novel event—the arrival of an escaped circus elephant in the garden, or your parents dressing up as Louis XVI and Marie Antoinette for a fancy-dress party—may well be quickly forgotten because it doesn't fit any script (unless they go every week to fancy-dress parties dressed as historical characters). And any individual event in a scripted routine will be forgotten because it has become assimilated into a script and can't be reconstructed as a separate event.

Clearly, Emily's working memory is fine. The events of every day are retained in some detail as she attempts to make sense of the world, although she doesn't always succeed. One evening, her father tries to explain that he is going to buy 'what is called an intercom system that we plug in, Stephen, into Stephen's room . . . When Stephen wakes up we'll hear him cry.' Emily's monologue after her parents leave the room suggests that she believes an intercom is something you plug into a baby to make him cry.

Interestingly, in the light of Madeline Eacott's work on people's memories of the birth of a sibling, although Emily's baby brother Stephen was born during the period of the recordings, she made no reference to the event in her monologues during the week it happened.

All of this is leading up to Nelson's analysis of childhood amnesia and why it might occur. For something to be remembered it must either be rehearsed, perhaps by repeated discussion with someone else, or cued in some way, by a question or by coming across a reminder of some specific element that was unique to that particular meal, shopping trip, or other scripted event. Without such triggers, all these memories become inaccessible as the child grows older. The reason is simple: they have done their job in enabling the child to develop expectations and behave effectively. The future is what matters, not the past. So back to Madeline Eacott's experiment with her own children. The new 'game' with the weird shapes and nonsense names was around for a few weeks and then disappeared without trace. It didn't fit any scripts that the children might already have had; it wasn't rehearsed, by repetition, or cued, by the shapes being brought out a few weeks later, so, as Nelson's theory would predict, it disappeared entirely from the children's memories.

In 2007 Emily Oster told me that she remembers nothing of the taping of her crib conversations, but on reading the book in which her conversations are dissected in meticulous detail, there was much that she recognized about herself:

Most of us get the opportunity to see what we *looked* like as a child, but few of us can hear so clearly what we were thinking. And just as I can see the resemblance to my adult self in old photographs, I hear a lot of myself in these transcripts. I still like to plan the events of the next day before I go to sleep, and I still talk to myself with surprising frequency (although my husband

Jesse notes that I now usually pretend I'm talking to him). Some of the phrases I used back then are too good to give up—for example, I often find myself saying, 'and won't that be funny!'

A few of the psychologists in the book hypothesize that the monologues were my way of understanding the puzzles of my world. I'm not sure if that was true at the time, but it's certainly true now. Sometimes closing the door to my office and talking to myself is the only way to get through a difficult research problem.[28]

The uniqueness of the transcripts of Emily's dialogues and monologues is both a blessing and a curse: a blessing, because there has been only one other, less comprehensive, attempt to obtain such material, and a curse, because it can be risky to generalize from the behaviour of one individual to a theory about all humans. And Emily is herself an unusual individual. Bright and precocious, with advanced verbal skills for her age, she has gone on to become, even in her twenties, an economist who produces original ideas and presents them in a fluent way.*

So Katherine Nelson's theory for childhood amnesia is that all the things that happen to a child during that period are perceived and analysed—to the best of the child's ability—and then incorporated into a general picture of the world, a kind of competence for living. At this point, the process of perceiving, and remembering for long enough to improve competence, has done its job and the individual elements, the events the child has experienced, are no longer necessary to retain as separate memories.

But this raises a related question that Nelson turned her mind to. However little remains of our earliest years, we all do retain some memories from childhood. So what's the explanation for those memories that survive the period of childhood amnesia?

For Nelson, the crucial difference between what survives and what doesn't is sharing. As we develop speech as children, we get enjoyment and social approval from sharing memories with others. What's more, if those others are our parents, they use the dialogue to do much more

* For an example of how well she speaks nowadays, see http://www.ted.com/index. php/talks/view/id/143

than just listen and nod approvingly. There are mechanisms at work in a conversation between a parent and a child which shape that child's entire autobiographical identity.

'The suggestion here,' says Nelson, 'is that this activity is learned in early childhood, and the result of this learning is the establishment of a store of memories that are shareable and ultimately reviewable by the individual, forming a personal history that has its own value, independent of the general memory function of prediction and preparation for future events.'[29]

Robyn Fivush, an American psychologist in Atlanta, has for the last twenty years investigated the way conversations between parent and child help to select, reinforce, and combine individual childhood memories into a coherent sense of autobiographical identity.

I met her where she works, at Emory University—an interestingly-named place for a memory researcher to work—and she told me of her studies of how parents and children talk to each other:

The kind of memory I'm interested in is family reminiscing where families just talk about the past together—'Oh that was like the time we went to the zoo—remember you saw the giraffe.' There are many things going on in these conversations. One is, if you went to the zoo what animals did you see? That's facts about the world, facts about the event—you either did or did not see the giraffe. Did you have fun? Well, maybe mom had fun and the kid didn't, or maybe brother Billy had fun but sister Susie was scared. So you have the evaluations and interpretations. A lot of what goes on in family reminiscing is about this evaluative aspect of it. There is a lot of talk about emotion and 'wasn't that fun and don't you love Grandma? . . .' So these conversations are focussing on what we as a culture deem important about this memory—'what's important about the zoo?' It's seeing the animals and what the animals were doing, so let's focus on that, so there's a highlighting or focussing that goes on in these conversations. It doesn't mean the memory becomes inaccurate, it means the child becomes focussed on 'what's the culturally important thing to talk about, what makes things to tell somebody about? What's interesting?'

In Fivush's view, these kinds of memories become valued for themselves. We cherish them, go over them in our minds, repeat them to others, and this process helps to pin down the person we are, the kind of person who goes to the zoo, sees a giraffe, and has fun. We must be that kind of person because our mother has focused on just those aspects of the experience. At the time, we might have been fascinated by seeing a gold-coloured cigarette packet on the path, or a leaf partly eaten away by insects, but that's not what parents pay admission money to zoos for us to see, and so any incipient curiosity might have been dismissed early on in the conversation.

Fivush and her colleagues recorded many dialogues between parents and children of differing social classes and cultures, looking for the shaping role of parental style on childhood memory. Here's an extract from one of the dialogues:

MOTHER: Do you remember what Auntie Elizabeth brought you that day when you saw your brother for the very first time? Do you remember what she brought you?

CHILD: Yeah.

M: Your most favorite thing in the whole world? Your favorite friend? Who is it?

C: Baby Dillon (her brother).

M: Baby Dillon. Is that who your most favorite friend is?

C: Yeah.

M: That's real special. I was talking about something that's furry and brown and real soft.

C: Yeah.

M: It was the first time you got your . . . ?

C: Yeah.

M: Who do you sleep with every night?

C: Teddy bear.

M: That's right.[30]

This is particularly interesting because it relates to one of the questions Neisser and Usher, and then Madeline Eacott, asked subjects about their early memories connected with the birth of a sibling.

There's no evidence here that the child has a strong memory for her teddy bear as a present from Aunt Elizabeth. In fact, it requires an irrelevant question—virtually a statement—from the mother to make that association. She might just as well have said 'The teddy bear you sleep with every night was actually given to you by Aunt Elizabeth when your baby brother was born.' But the conversation has helped to turn that fact into a memory for the child. It happens to be a 'true' memory but may not be an accurate one. It's true (unless the mother is lying in her teeth, or of course, has a bad memory herself) but may not be accurate because the child doesn't appear to be able to summon up for herself an actual memory of the handing over of the teddy bear at the time of her new brother's birth.

Richard McNally, a Harvard psychologist, comments:

Autobiographical recollection is a reconstructive, not a reproductive, process. When we remember an event from our past, we reconstruct it from encoded elements distributed throughout the brain. There are very few instances in which remembering resembles reproducing. These include reciting poems, prayers, telephone numbers, and other material memorized by rote.[31]

One consequence of the role of parental conversation in constructing memory is that the type of memories we retain depends to a certain extent on the conversational style of parents but also on the personality of the child.

'There are individual differences in how interested kids are in engaging in these conversations', Robyn Fivush told me:

Some kids are very 'Oh yeah, let's sit down, let's do this. Tell me more', and they participate. Other kids not so much. So it's not just the parents. Mothers generally talk more to their kids than fathers do. And they sit and reminisce with their kids more than fathers do. There are also individual differences, so some mothers and fathers when they talk about the past are what we've been calling highly elaborative: they tell rich, embellished stories, they talk a lot about the emotional aspects—'Wasn't that fun?' 'Oh, what do you remember about that? That's right...', positively evaluating the child's contributions. Some parents are less elaborative. They seem to see these conversations more as memory training—'OK, so we went to the zoo—tell me what animals we saw.'

Fivush has also found gender differences in the way parents help children form memories:

45

Parents seems to be more elaborative with girls than boys. By the end of the preschool years girls are telling longer, more detailed stories of their own personal experiences than boys are. They include more emotion and they seem to include more general evaluation.

What we know is that, in this sort of mostly white Western population, mothers who are highly elaborative when their kids are very young are highly elaborative through middle childhood. And the kids of these more elaborative moms come to tell more richly detailed stories of their own lives. Do they remember these things differently? I think they probably do.

Fivush believes that the style of reminiscing affects the content of the memory, and that parent–child conversations lead to more than just certain things being remembered and others forgotten. They shape a child's perception of what is important about living in a social world—relationships, gender, materialism, a love of words, values, and ethics. In a word—identity.

There is some suggestion now that mothers who are more elaborative have kids who have a more stable and more differentiated sense of self—'I know I'm good at this but I'm not so good at that.' In early childhood, through seven or eight, when people say 'sense of self' they usually refer to the ability to talk about my characteristics, beyond 'I have yellow hair and my mommy's name is Betty', to be able to say 'I really like to play with blocks but not so much with dolls', so it's still pretty concrete but it's an ability to know what you're good at, what you like, what you don't like, to have a more complex understanding of differentiating between my physical traits, my social abilities, my psychological traits.

The data underlying this 'social interaction hypothesis', as it's called, of Fivush and Nelson and others, came initially from studying children in a Western society. But child-rearing practices and parent–child interactions are very different in other societies. Could different styles of interaction lead to different types of memory? Could we look at what people from Eastern cultures remember, say, and see if there are differences that are related to different styles of child-rearing?

Qi Wang, a Chinese-American researcher at Cornell University, believes we can. She compared the early memories of two groups of psychology students, one in the US and one in Taiwan. They were given the word cues 'Self', 'Mother', 'Family', 'Friend', and 'Surroundings' to trigger early memories in each of those areas. There were significant differences between the two groups, which Wang related to the different cultural backgrounds of the students:

Euro-Americans frequently reported memories of specific, one-moment-in-time events and focused on their own roles and autonomy, even when recalling memories about mother and family. In contrast, Taiwanese often described early experiences of generic, routine events with a salient social orientation.

Wang gives examples from the results of the survey:

Self:
US: Preschool learning to tie my shoes on my own and playing with all those 'skills games' with shapes and hand-eye coordination. I got a gold star and realized my friend didn't.

Taiwan: Ever since I was little, my relatives would compare me with my girl cousin. Everyone said that I was the more sensible one. My girl cousin and I were 4 months apart in age.

Mother:
US: My mother dressed me up as a queen for Purim at my preschool. She let me wear all her fake jewelry—very long beads & makeup. I wore a long white shirt as a dress with a belt & a paper crown. I was very happy when she walked into school with me.

Taiwan: My mom was always smiling. She was always working, and didn't have time to talk to me or listen to me. Even in my memory she was always busy walking around, and never stopped to take a rest.

Family:
US: Going to Niagara Falls on a family trip. I was young so I still asked my parents if I got the shampoo out of my hair. My dad said I did. However, when we walked behind the falls in caves, my hair started to bubble because I didn't get it all out.

Taiwan: Grandpa, grandma, dad, mom, my little brother, and our family's puppy. The whole family was having breakfast in the dining room one morning. We had congee* that day. I still remember the dishes: shredded pork, clam . . .

* A type of rice porridge that is eaten in many Asian countries.

Friend:

US: In second grade, my best friend moved on my birthday. She called me on the phone to say goodbye, and I cried. My mother still made me practice the piano that night, and that made me angry and more upset.

Taiwan: In kindergarten, I had a pretty good relationship with a little girl who lived near my neighbor. Her family raised ducks for a living. Often, after school, we would play in the field behind her house or play hide and go seek in her house.

Surroundings:

US: 3 years old, lying on my 'comfort blanket.' The room had a carpet then (sky blue). My crib had all sorts of stuffed animals. It was mid afternoon and raining. My brother and I were playing with blocks.

Taiwan: In the park, there were four or five different recycle bins. There were mothers and their children taking a walk and chatting. I was just about to go to preschool.[32]

Interestingly, the average number of times the American subjects in the sample above used 'I' or 'me' was 3.6 per answer, whereas for the Taiwanese it was 1.6.

This cultural difference was also reflected in the age at which the events recalled by the two groups occurred. For each cue word, the mean age of first memory in years and months is given for each group:

Self: US 4:7, Taiwan 5:10; **Mother**: US 3:10, Taiwan 5:5; **Family**: US 4:6, Taiwan 5:10; **Friend**: US 5:0, Taiwan 6:4; **Surroundings:** US 4:4, Taiwan 5:6.

A survey of the earliest memories of Koreans found that they were an average of sixteen months later than the Chinese, who were in turn six months later than white Americans.

In her discussion of the findings, Wang wrote:

An emphasis on individuality in Euro-American culture may drive the early emergence of a personal event memory system that serves to construct the distinctiveness of the individual, whereas an emphasis on interconnectedness in Chinese culture may direct cognition to social knowledge for maintaining existing social orders and conventions, rather than to the early development of personal event memories.[33]

Wang thinks this difference may be due to the perceived importance of personal remembering across the two cultures. If there are fewer questions along the lines of 'What did you do today?' and 'Do you remember what Aunt Elizabeth gave you?' and 'How did you feel when that boy hit you?' there are likely to be fewer long-term personal memories created.

Other research has found similar results using different methods, including an analysis of written autobiographies:

In contrast to this all-dominating self-focus of Western autobiographical accounts, people from Asian cultures tend to provide brief accounts of past experiences that give a great emphasis on social relations and moral rectitude while they show little concern with the positioning of individual roles, preferences, and feelings.[34]

Once the idea of cultural differences in autobiographical memories emerged, wherever scientists looked they found evidence that supported the theory.

Two studies showed that the earliest dreams that Westerners remembered occurred on average 14 months earlier than Asians. The researchers suggested that this was because 'Compared to Asians, Caucasians reported talking more frequently with parents about their dreams in childhood, receiving stronger parental encouragement to share dreams, and feeling more comfortable doing so'.[35]

David Pillemer reported on interesting differences in personal storytelling among cultures. His dichotomy, which roughly matched the East–West division of Wang's work, was between what he called Individualistic and Collectivist cultures:

It is not uncommon for people from individualistic cultures to freely recount personal anecdotes and even intimate details of their private lives; they appear to have the need to express [their] own thoughts, feelings, and actions to others. In contrast, free-flowing conversations about the personal past may occur less frequently in collectivistic cultures. Collectivists often communicate indirectly or nonverbally, under the implicit assumption that other people will know or can infer their intentions. Even in psychotherapy sessions, Japanese patients expect the therapist to understand their private selves without probing: 'For the therapist to ask questions of the patient is considered to be intrusive at best and insulting at worst'. Individualists pursue intimacy by exchanging detailed personal information; collectivists achieve intimacy through implicit, empathic understanding.[36]

So we've arrived at a number of factors which contribute to the nature and content of our earliest memories. It seems that, while there may be no way to be sure which are accurate and which are not, it is likely that the earliest memories, particularly if they are based on fleeting images or sounds, and provided we do not claim that they happened during the first few months of life, could well be genuine. On the other hand, *narrative* memories from early on—little stories about who did what to whom—are unlikely to be pure unmediated memory. They are much more likely to be as a result of discussion at the time with a parent or sibling, perhaps merely a reminder from the older person along the lines of 'Do you remember telling me about X or Y or Z?' The teddy bear example above is typical. So while many of us are confident that we are remembering accurately our earliest life events because we can't remember ever discussing them with our parents, those discussions themselves might well have been forgotten, perhaps because we incorporated them into a 'script' of some regular parent–child discussion which was then assimilated and therefore forgotten, at least as a stand-alone event memory.

As for why we have forgotten so much, there are several reasons. Entirely novel events were forgotten because we couldn't fit them into what we were learning about the routines of life that we needed to understand in order to adapt our behaviours to everyday life. *Quite* novel ideas may have been remembered for a while, so as to incorporate them into scripts if they occurred again. And the scripts themselves became part of our regular behaviour in a way that made them inaccessible to normal recall.

It is only as we get older, become more verbally skilled, and need to play a more social role in life that our memories become important for different reasons. Now, exchanging memories becomes a social currency. 'You tell me your story and I'll tell you mine.' And that sociability starts in the home. The things we remember, in order to tell our mother or father or brother or sister, are the things they lead us to believe they will be interested in. And how do we know what they will be interested in? Because they teach us by the way they converse, shaping our stories, and our memories.

None of this is to suggest that childhood memory—or indeed adult memory—is well understood. There are still many puzzles about what children remember, and what adults remember about their childhood. The murkiness, inaccuracy, and randomness that applies to the things children remember seems to operate among adults as well. A simple view of memory might say that on the whole, as we grow older the main changes in what we remember are due to a process of attrition. But the reconstruction of memory seems to go on throughout life. Daniel Schacter, in his book *How the Mind Forgets and Remembers*, describes an experiment that shows remarkable and unexpected changes in the memories of adults over time:

We're all capable of distorting our pasts. Think back to your first year in high school and try to answer the following questions: Did your parents encourage you to be active in sports? Was religion helpful to you? Did you receive physical punishment as discipline? The Northwestern University psychiatrist Daniel Offer and his collaborators put these and related questions to sixty-seven men in their late forties. Their answers are especially interesting because Offer had asked the same men the same questions during freshman year in high school, thirty-four years earlier.

The men's memories of their adolescent lives bore little relationship to what they had reported as high school freshmen. Fewer than 40 percent of the men recalled parental encouragement to be active in sports; some 60 percent had reported such encouragement as adolescents. Barely one-quarter recalled that religion was helpful, but nearly 70 percent had said that it was when they were adolescents. And though only one-third of the adults recalled receiving physical punishment decades earlier, as adolescents nearly 90 percent had answered the question affirmatively.[37]

It seems that for these men—and probably for all of us—the ability to remember accurately *anything* about our personal past can sometimes be no greater than chance.

51

This, of course, is not good news for people caught up in situations where the question of whether a memory—particularly a distant one—is accurate or not can lead to psychological illness, accusations of abuse, or even criminal convictions. Perhaps a closer study of the details of how we use our memories in everyday life can help us understand what memory is for and when we can rely on it.

4

....

Reconstruction

'Please assume that there is in our souls a block of wax,' Socrates said,
'in one case larger, in another smaller, in one case the wax is purer, in
another more impure and harder, in some cases softer, and in some of
proper quality... Let us, then, say that this is the gift of Memory, the
mother of the Muses, and that whenever we wish to remember
anything we see or hear or think of in our own minds, we hold this
wax under the perceptions and thoughts and imprint them upon it,
just as we make impressions from seal rings; and whatever is imprinted
we remember and know as long as its image lasts, but whatever is
rubbed out or cannot be imprinted we forget and do not know'.[38]

What Socrates didn't know about memory would fill the Parthenon,
but the 'block of wax' metaphor, or updated versions of it depending on
current technology, was actually a central feature of beliefs of how
memory works until the middle of the twentieth century, in spite of
the fact that several classic experiments were carried out in Cambridge
during the First World War which showed how creative, constructive,
and inaccurate many of our memories are.

The experimenter was a thirty-year-old university lecturer, Frederic
Bartlett, who was interested in anthropology but decided to acquire
some skills in experimental psychology for possible use in anthropo-
logical work. In fact, the research he did during and after the War made
a major contribution to the science of memory, and he continued to
work on psychology for the rest of his career.

Looking back now at the details of his experiments, they seem
surprisingly simple for the wide-ranging deductions that were made

from their results. As he said himself, anyone could have done them, and indeed could do them today.

What Bartlett did was to recruit subjects, not always psychology students, from the young men and women of Cambridge, and present them with a passage to read. He then tested them at intervals to see what they had remembered—and forgotten—of the material in order to draw conclusions about how we remember. And the use of the verb is very important. Bartlett always said that he was investigating 'remembering' not 'the memory', and he believed—and his results showed—that remembering was a process that owed little to Socrates' 'block of wax' and other people's 'memory traces'.

When I studied undergraduate psychology at Cambridge some decades later, it was standard for the lecturer to repeat Bartlett's most famous experiment, by asking the students to listen to the passage that Bartlett used with twenty or so subjects. It was a folk story called *The War of the Ghosts*, and when starting to write this book I tried to remember anything about it after forty years. Only one element remained in my memory—'something black' came out of someone's mouth.

Bartlett's experiments showed, unsurprisingly, that as time passed people remembered less and less of the stories he used as stimulus material, but what they *did* remember was shaped by personal factors such as attitude, temperament, interests, and what he called 'effort after meaning'.

The War of the Ghosts is a North American folk-tale, chosen because it was culturally foreign to his subjects and would therefore present certain unpredictable or even incomprehensible elements. Bartlett wanted to see if the changes in the story as it was recalled at intervals would reflect one of the factors he believed was at work in all remembering, a process of reconstruction based on previous experiences and memories, and on attempts to make sense of the incomprehensible.

One night, (the story began,) two young men from Egulac went down to the river to hunt seals, and while they were there it became foggy and calm. Then they heard war-cries, and they thought: 'Maybe this is a war-party'. They escaped to the shore, and hid behind a log. Now canoes came up, and they heard the noise of paddles, and saw one canoe coming up to them.

The young men are invited to go and make war 'on the other side of Kalama' by five men in the canoe, and one goes with them while the other returns home.

Back home at Egulac, the young man went ashore to his house, and made a fire. And he told everybody and said: 'Behold I accompanied the ghosts, and we went to fight. Many of our fellows were killed, and many of those who attacked us were killed. They said I was hit, and I did not feel sick.'

He told it all, and then he became quiet. When the sun rose he fell down. Something black came out of his mouth. His face became contorted. The people jumped up and cried. He was dead.

The full story was given to twenty subjects to read. The subjects read the story through to themselves twice, at their normal reading rate, and then they were tested at irregular intervals, 'as opportunity offered' Bartlett said. It has been suggested that this meant as and when he bumped into the subjects in the Cambridge streets. In his research papers, Bartlett reprints examples of the many iterations of the story from individual subjects. Here's the fifteenth remembering of one person, trying to recall the story after two and a half years:

Some warriors went to wage war against the ghosts. They fought all day and one of their number was wounded.

They returned home in the evening, bearing their sick comrade. As the day drew to a close, he became rapidly worse and the villagers came round him. At sunset he sighed: something black came out of his mouth. He was dead.[39]

As I myself had experienced, elements of this folktale disappeared very quickly, while others, the odder ones, stuck in the memory. The story goes that one day Bartlett was cycling along King's Parade in Cambridge when one of his former students spotted him as she was walking along. She was puzzled that the words 'Egulac' and

'Kalama' came into her head for no apparent reason, but realized that they were remnants of the folktale she had been told to read. Two years later those words were still in her memory, and when pushed to remember more she managed to come up with an incident or two which bore a vague resemblance to parts of the original story, but this was probably reconstruction rather than specific memory.

Bartlett was interested in memory for other types of material, in addition to the unpredictable elements of an exotic folk-tale. Clearly, the very foreignness of the story meant that certain elements stuck out as memorable. But what about more culturally familiar information? What does our memory preserve of such passages? And this isn't just to do with remembering facts. To see how people remembered the ordered steps in an argument, Bartlett used a different passage as his initiating story, a paragraph about evolution:

Modification of Species

One objection to the views of those who, like Mr Gulick, believe isolation itself to be a cause of modification of species deserves attention, namely, the entire absence of change where, if this were a *vera causa*, we should expect to find it. In Ireland we have an excellent test case, for we know that it has been separated from Britain since the end of the glacial epoch, certainly many thousand years. Yet hardly one of its mammals, reptiles or land molluscs, has undergone the slightest change, even though there is certainly a distinct difference of environment, both inorganic and organic . . . (The paragraph continued for another 80 or so words.)[40]

With this passage Bartlett used a different method of transmission. Like the children's game of Chinese Whispers, the passage was read by a subject who then passed on his memory of it to another, who passed his memory of it to a third, and so on. Here's what remained of this passage after ten transmissions:

Mr Garlick says that isolation is the result of modification. This is the reason that snakes and reptiles are not found in Ireland.

Almost every word of this summary represents an inaccuracy in remembering.

Using this method with a passage about cricket, presented to a group of public schoolboys who knew and were enthusiastic about the sport, Bartlett gave the following verdict on how well the remembering process was carried out:

It will be seen at once that in this series almost every possible error has been made.... The title soon falls out; all the proper names but one disappear, and that one is assigned to the wrong eleven and has no place in the original. Robinson bats first well and then badly; the bowling is first good, then easy and then good again. All the numbers go wrong or else drop out completely. The sides are transposed. About the only new element introduced into the story, except the name Robinson, is the change to the first person singular, and the statement eventually that Middlesex were lucky to make a good score.[41]

There is a homespun quality about this work: no abstruse scientific theory, no complicated experimental apparatus, and research reports that are easy to read and sound almost like common sense. These were all a reflection of Bartlett's own 'man on a bike' personality. He wasn't someone who thought that the science of psychology was only valid if it produced complex and erudite results. The discoveries that emerge from the scientific pursuit of psychology have often been criticized because they sometimes just confirm 'common sense.' With physics, we have very little perceptual awareness of atoms, say, and our commonsense views are unlikely to be of much use in understanding them. But with memory, where the commonsense view is that memory is like some kind of recording, leaving 'traces' in the brain, Bartlett showed how good psychological experimentation could lead to a more accurate understanding of memory.

There had been a tendency in the late nineteenth century, when psychology became a different discipline from philosophy, for the study to become 'scientized'. Rather in the way physicists were breaking down matter into smaller and smaller elements, psychologists were trying to isolate the different factors that they believe operated independently of each other, and look at them one by one. A prime example of this was one of the earliest memory researchers, Hermann Ebbinghaus. In choosing what to present to his experimental subjects to memorize, he tried to avoid any stimuli that would arouse personal associations in the subject's mind, because this would mean that he couldn't be sure that the subjects of his experiments would start from the same baseline. He therefore devised lists of nonsense syllables, which, in his view, would mean nothing to anyone, and used them to test memory.

Here is a selection of the syllables he used: DAR, BEL, FOT, MUK, LIM, DOR, HAK, KOD, JIH, BAZ, FUB, YOX, SUJ, XIR, DAX, LEQ, VUM, PID, KEL, WAV, TUV, ZOF, GEK, HIW.

Ebbinghaus made 169 separate lists of such nonsense syllables and then, with himself as his own subject, relearned each list after a different interval, from 21 minutes to 31 days. He used the time it took him to relearn each list as a measure of 'forgetting' and plotted the results on a curve which showed a sharp drop-off at first and then a shallower decline until the memory for a list stabilized at about 20 per cent of the learnt material. He also presented tables of statistics to show that the results had general significance.

Bartlett's approach was the opposite to Ebbinghaus's. He believed that using significance-free stimuli was both pointless and actually unachievable:

Any psychologist who has used them in the laboratory knows perfectly well that lists of nonsense syllables set up a mass of associations which may be very much more odd, and may vary more from person to person, than those aroused by common language with its conventional meaning.[42]

A glance down an Ebbinghaus list will probably generate semantic associations for all of us, and not necessarily the same ones. Admittedly, I am applying English associations to German nonsense, but KOD and HAK are fishy, BAZ reminds me of someone famous, DAR is Arabic for 'house', MUK is dirty, BEL rings. Even when I consider what is real nonsense in English—JIH, FUB, XIR, GEK, and so on—my mind tries to create significance for them to make them easy to remember.

Supporters of Ebbinghaus's method might say that with routine, uniform exposure to the syllables the subjects would dispense with these associations and concentrate purely on learning them as meaningless sounds or groups of letters. In that case, Bartlett wrote, 'the remedy is at least as bad as the disease. It means that the results of nonsense syllable experiments begin to be significant only when very special habits of reception and repetition have been set up. They may then, indeed, throw some light upon the mode of establishment and the control of such habits, but it is at least doubtful whether they can help us to see how, in general, memory reactions are determined'.[43] In other words, using nonsense syllables as stimuli teaches us only about how we remember and forget nonsense syllables.

Bartlett was interested in the richness of remembering, and how we remember the kind of material that is part of our everyday lives, to do with events and people and emotions and surroundings and so on. It's like the difference between chemistry and biology. Chemists study atoms and the way they combine into molecules; biology, which is still dependent on the activities of atoms and molecules, looks at much higher-level aggregations of chemical units, so that the way hydrogen and oxygen atoms combine to form water doesn't really help us understand the effects of dehydration on the body's physiology.

For Bartlett, the remembering process involved a whole chain of cognitive events over time, starting with the first exposure to an event or scene—perception:

Nobody who reflects upon how variously determined are the processes and content of perception will be prone to give a careless allegiance to the theory of lifeless, fixed and unchangeable memory traces.[44]

Using inkblots, he showed how even a basic monochromatic shape could trigger a bewildering variety of perceptions in different subjects, each of which would then determine how it was remembered. He gave the following list of associations triggered by the same inkblot in nine different people:

Irate lady talking to a man in an arm-chair; and a crutch.

Bear's head, and a hen looking at her reflection in the water.

Angry beadle ejecting an intruding beaver which has left footmarks on the floor.

A man kicking a football.

Lakes and green patches of meadow-land.

Scarecrow behind a young tree.

Tiny partridges newly hatched.

Animal pictures and the Crown Prince of Germany.

Smoke going up.

Such first impressions often shape the later process of recall. The choice of what real life concept to impose on the ink-blot, for example, is related to the subject's attitudes, preferences, interests and temperament. No one can reproduce the shape of a random inkblot without associating it with some real object—unless he or she has what is called eidetic memory, in which every tiny detail of a visual impression is preserved—and inexorably, when trying to remember later what one has seen, to a certain extent the memory takes on more of the shape and structure of the content that was imposed on it when first seen.

Here's a story I stumbled on that illustrates the role of the observer's pre-existing psychological state in perceiving and then remembering. It was in an autobiography I was reading for a different purpose, of a man called Eric Siepmann:

I used to go for walks among the Maures mountains, just above Toulon . . . One day I reached a village, round the bend of a rock path; and I was attacked by a flock of angry geese. Bending their necks, and hissing through their beaks, they gave me a fright, because my thoughts had been far away from the stony path, which they regarded as their own. I dodged them, with my back to a wall. Having got ahead of them, I picked up a few pebbles and tossed them at the geese to discourage them from following me. Then I walked past a farm, looking at it out of the corner of my eye as I felt self-conscious about my encounter with the geese. In the courtyard of the farm was a clothes-line, and a girl was stretching her arms to hang up a dress. The poise of her arms was graceful, and she was wearing a pink dress.

For some reason my blood thudded in my temples, as I hurried away. Was this the girl for me? She lingered in my mind. After much misery, I was filled with an abnormal happiness.

Here was a perception that was obviously destined to stay with the author as a result of his emotional state, yearning for female companionship, or, as Bartlett wrote, 'some preformed bias, interest, or some persistent temperamental factors'.

The memory of the girl would have lingered, perhaps become stronger, and been recalled from time to time, had Siepmann not returned, in the hope of seeing her again.

'Two days later,' Siepmann continued, 'I walked in the direction of the same village. The cork-trees danced, as usual. The geese were there, but I walked proudly past them. In the courtyard of the farm, the clothes-line fluttered. Suddenly, I realized what had happened. There had been no girl. What I had seen, out of the corner of my eye, was a pink dress, inflated by the wind. Perhaps I am the only man in history who has fallen sincerely in love with a piece of laundry. But my imagination had been inflamed. For two days, I had been happy'.[45]

This illustrates very well a key principle that Bartlett believed emerged from his research: 'It is fitting to speak of every human cognitive reaction,' he wrote, '—perceiving, imaging, remembering, thinking and reasoning—as an *effort after meaning*'.[46]

Each of the stories or passages Bartlett used as a test of remembering changed with every repetition in ways which confirmed his theories about the way we remember things. Far from there being a chronological series of accurate images and sounds from the past which we riffle through when we are trying to remember, our cognitive processes continually change new material at the point of perception and afterwards in the light of the pre-existing memories of previous experiences. If Eric Siepmann had been a Trobriand Islander and never seen a woman in a dress, he would not have linked his perception of laundry to a previous memory of a woman. Perhaps, instead, he would have compared it—instantaneously—with a previous memory of a sail in the wind, and reconstructed a boat in his mind instead of a woman.

'What we do,' Bartlett wrote, 'is to go on building up all the time, throughout the whole of our waking life, a mass of more or less organized experience, the items of which are always liable to affect one another and change one another in many ways. So if we speak of "traces" in the mind which are used when we remember, we have got to think of them rather like lesson material that is being constantly revised, and each time it is revised new material is apt to be added and old material dropped out or altered'.[47]

And this 'lesson material' is organized in a very personal way, as we grow and develop. It's as if we had a set of 'exercise books' organized by the topics that interest us or are important in some way to the unique personality that each of us has. When we need to consult the past we do not have to sort through a mass of unconnected sights and sounds and feelings, but are directed by the needs of the present to the right area of memory to consult the past. As Bartlett's work showed, separate memories are easier to retain when they are organized into stories with narrative and structure, and this inevitably leads to some elements dropping out, being 'forgotten'. And it is *reconstructions* of the past we consult, that have become revised over the years, in ways that we are rarely aware of as we re-remember them.

For small children, this revision process is what takes place typically during conversations with parents, which not only lead to revision of the memories themselves but also of the scale of importance each memory should be granted. Just like Emily's bedtime monologues and conversations, but self-reflectively as well, our acts of perception and recall work together to process our experiences continually and provide useful lessons for the present and the future.

One result of this method of processing is that strict chronology does not usually matter. This means that our memories do not seem to have obvious 'date-stamps' on them, as they might do if remembering was 'trace retrieval' from some kind of databank rather than a process of reconstruction.

But there have been psychologists who believe that memories do have some kind of built-in dating information. One theory suggests a 'time tag' attached to each memory; another, that the time that has passed since an event occurred can be gauged by the strength of the memory of that event; a third theory suggests that each event is associated with a position in the sequence of events that happen to an individual, like bags on a moving conveyor belt.

A little introspection might suggest that we don't actually use any of these methods to date our memories, at least consciously. Regardless of whether they are accurate, memories 'look' or 'feel' like representations of events from our past, but they are not usually accompanied by a specific date or time. Nor do they have an intensity scale attached to them, such that recent events are vivid and more distant ones fainter, in some mathematically predictable way.

But introspection has been a dirty word for some psychologists, who prefer to have hard evidence. One major experiment which throws light on the 'datability' of memories was carried out in a fifteen-year-long project by a group of psychologists at Kansas State University. Instead of the kind of retrospective work like that of Madeline Eacott, for example—'Can you remember when X happened?'—the psychologists attempted to gather material about events in the lives of students as they happened, by asking the subjects to keep diaries over a period of several weeks, and then at a later date investigating which events the students remembered and how well they remembered them. This is known as a prospective study. The subjects were asked to keep a note each day of a memorable event and of another event they thought was less memorable. These diaries were then collected at regular intervals so that the participants didn't refresh their memories by rereading them. At a later date they were reminded of particular events they had written down and asked questions about their memory for those events.

These are the kind of memories that were written in the diaries:

Anne and I watched Frank compete in the Little Apple triathalon.

First chemistry lab of the semester. I didn't get everything done.

Highlighted Linda's hair while she talked on the phone.

I had surgery on my foot at Memorial Hospital.

Received a phone call about my dad's death.

We had our first cross-country meet at Iowa State. It was on a golf course and golfers played through!

Went over to Tim and Brian's to see their new couch.

Went to an all-campus party at Phi Delta.

While talking on the phone to the cable company, the cable came back on the air. I was so embarrassed![48]

When the psychologists asked the diarists later to date the events, they found that some were dated exactly, some moderately well and some very poorly. The accuracy of dating depended on the *content* of the memory rather than on any additional characteristic such as a 'time tag', intensity of the memory, or a position in some kind of ranked list of memories.

The best-dated in this short list were: the death of the father, the cross-country meet, and the foot surgery. This was put down to strong associations with a schedule or with a specific date that was likely to be remembered for its own sake because of the unusually significant nature of the event.

The moderately well-dated memories were: watching Frank compete, the all-campus party, and the first chemistry lab of the semester. It was suggested that these were not unusual enough to be firmly dated and yet they stood out enough to be dated with an error of less than seven days.

The memories that were very difficult for the subjects to date with any degree of accuracy were: highlighting Linda's hair, going to see Tim and Brian's new couch, and the cable company phone call. The experimenters said that these 'seem to be drawn from the events in a person's life that are not likely to be at all tied to ideas about time'.

Interestingly, the ability to date a memory does not seem to be connected with the significance of the memory to the subject. One of the least well-dated memories was actually the rather embarrassing incident with the cable company, which the subject obviously remembered quite vividly even though he or she couldn't put a date to it.

One finding came out of an analysis of the errors made when the subjects tried to recall the date of a specific memory. Many of the diary events were recalled as happening about seven or fourteen or twenty-one days before or after they happened, suggesting that some aspect of the experience was assigned to a part of the week—a day, a weekend, midweek—rather than a specific date or after a certain lapse of time.

If the function of remembering is to provide templates for us to consult as we go on in life, then exactly when something happened is often not a particularly useful piece of information. In fact, it could actually be counterproductive if we were slaves to chronology in the way our memories were arranged.

With the huge amount of sensory data that we process every day, we often encounter new situations which we wish to compare with our earlier experiences. In these cases, the least useful way of sorting through all the data would be to go backwards chronologically until we found what we were looking for. We can access memories from decades ago as quickly as remembering an event that happened yesterday.

Imagine if, whenever we bumped into someone whose face we last saw five years ago, we had to consult, one by one, all the faces we had seen in the interim in reverse date order.

For someone who assembled a couple of dozen people and got them to read stories and look at pictures, Frederic Bartlett made an unexpectedly important contribution to our understanding of memory. His view of the reconstructive nature of remembering, formed in the 1920s and 1930s, underlies and is confirmed by the work of many modern psychologists. Roger Schank, the 'scripts' man, believes that storytelling plays an important part in all our lives and that the stories we tell, based on our memories, are every bit as liable to change as *The War of the Ghosts*:

In our desire to tell a story in the first place, we resort to certain standard story-telling devices. Those devices are part of our cultural norms for story telling and they reflect what is considered to be a coherent story in a culture. Since, in telling one's story to others, one wants to be coherent, one has to structure one's story according to these norms. This means, in effect, that one has to lie. Nothing in life naturally occurs as a culturally coherent story. In order to construct such a story we must leave out the details that don't fit, and invent some that make things work better. This process was seen in Bartlett's work on Eskimo* folk tales which were remembered by British subjects many years later as coherent stories while the original was certainly not coherent in a British context. This same process is at work when we tell our own stories. We tell what fits and leave out what does not. So, while our lives may not be coherent, our stories are. The danger here is that we may come to believe our own stories. When our stories become memories, and substitute for the actual events, this danger is quite real. We remember our stories and begin to believe them. In this way, stories shape memory profoundly.[49]

This work with adults suggests that there is a continuity between the way children remember and the way we all remember later in life. Once the period of childhood amnesia is over, children begin to understand the patterns of the world around them in a memorable way so that they can build upon and adapt them as they encounter new patterns. But they are also subject to the forces of change that occur when a memory is revisited. For the young child this occurs in his interactions with parents and siblings. For adults, the social setting plays a part, but our memories also change and evolve as a result of the way in which, in a changing world, we are constantly using our memories to help us make decisions. Bartlett uses the clumsy phrase 'turning around on our own schemata' to describe this process by which we search our past for items which are then used to satisfy current needs.

This process results in 'rationalisation, condensation, very often in a considerable rearrangement of temporal relations, in invention and in general in an exercise of constructive imagination to serve whatever

---

* Actually, *The War of the Ghosts* is a folk-tale of the native American tribes of the North-West Pacific coast.

are the operating interests at the time at which the "turning round" takes place'.[50]

Clearly, this means that *all* our memories—children's of their childhood, ours of our childhood, ours of what happened yesterday—are fallible. This is not a very helpful statement. What would be more helpful is some understanding of the specific parameters of this fallibility. Are there ways of distinguishing between accurate memories and inaccurate ones? Are certain things forgotten more readily than others? Is forgetting—the flip side of remembering—a process that can be studied and analysed? If we don't remember something that undeniably happened, where has the memory gone, and can it be recovered?

Over the last twenty years, many psychologists have tried to answer these questions, as childhood memory has become a battleground in what has been called 'one of the major mental health scandals of the twentieth century'.

Roger Schank wrote: 'We remember our stories and begin to believe them'. As we will see, it's when other people believe them too that the science of memory steps out of the laboratory into the dangerous world of the psychotherapist's office and even the courtroom.

5

· · · · · · · ·

Memory Wars Break Out

In 1990 George Franklin, a retired fireman living in California, was accused of the murder of an eight-year-old child, Susan Nason, twenty-one years beforehand. The case against him was based almost entirely on the evidence of his grown-up daughter, Eileen Franklin-Lipsker, who claimed that she had seen her father commit the murder at the time but had had no memory of the events for over twenty years. Eileen's evidence was given force by the fact that she described details that had apparently been known only to a small number of people directly involved in the investigation and were said not to have been published in newspapers or on television.

The Franklin case marked one of the earliest instances in a court of law of someone being convicted on the basis of an adult recounting a childhood memory that she had been unaware of for decades. It introduced a concept—the *repression* of childhood memories—that had barely been discussed or referred to before the 1980s. This idea was to be taken up during the 1990s and become enshrined in many people's minds as a basic tenet of the science of childhood memory. As the suggestion spread that adults could forget events of their childhood for decades and then retrieve them, that retrieval became the focus of techniques that people could carry out on their own or with the help of therapists. As one memory researcher sums up that period:

The 1990s were marred by a heated and often ugly debate concerning the accuracy of traumatic memories that had seemingly been forgotten for years or decades, only to be recovered in psychotherapy or in response to some triggering incident. Early discussions were divided sharply, with one side arguing that virtually all such memories are accurate and the other that virtually all are false. Although the bitter division has persisted, recent discussions have contended that both accurate and false recovered memories of childhood traumas exist, and have attempted to characterize the mechanisms responsible for each.[51]

One of the expert witnesses in the Franklin case was Elizabeth Loftus, who had become interested in the area of forgetting, repression, and false memories. Testifying in the case was to change her life and her career.

In October 2006 I spent the day with Loftus at her department at the University of California at Irvine. At nine in the morning, she strode down the corridor towards her office, smartly dressed and wearing a hat—she has a fondness for hats, and a large collection of them. It was her birthday and during the morning she was to receive birthday greetings and small gifts from students and colleagues. As she sat in a meeting she opened her mail. There was a birthday card from a mother who had been falsely accused of sexually abusing her child. There was also a letter from a prisoner, several handwritten pages starting 'I am serving a sentence for armed robbery and rape . . .'. 'That goes to Shari,' said Loftus. 'She handles all the prisoner letters'.

Loftus is a heroine to a large number of parents accused of sexual abuse of their children and the bane of people who believe in the concept of repressed memory of childhood events.

The waters are often muddied in discussions of this topic by imprecise definitions of the terms used. The concept of forgetting is quite a slippery concept. For example, if you don't think about something for a while, does it mean you've forgotten it?

'I think people can "not think" about something without necessarily forgetting it,' Elizabeth Loftus said. 'I don't like to think about the death of my mother unless I'm in the mood for dwelling on it and feeling sad for a while; if there's a reminder I can try to distract myself and think about something else'.

If you can't immediately recall what you did on your twenty-first birthday, does it mean you have forgotten it or repressed it? 'Ordinary' forgetting is usually something that is rectified by some trigger—a person you bump into who reminds you that you met ten years ago at someone's party, and then the whole incident springs into your mind. A 'repressed' memory would be something that is so inaccessible to consciousness that no amount of effort, or, indeed, no normal trigger would bring it back. And yet, the idea that traumatic memories can be deeply repressed, and, as some people allege, the more often the trauma is repeated the more the memory of it is repressed seems to stand right outside what science has established about memory and forgetting.

Normal forgetting and remembering are processes that follow certain laws, as Richard McNally points out:

First, ordinarily, repetition enhances memory for the class of repeated event. For example, the more often one flies in an aeroplane, the more likely one will remember having flown in aeroplanes. Even though memories of some flights might fade, or memories of some flights might get confused with other ones, one is very unlikely to completely forget that one has flown in aeroplanes. . . . Second, strong affect triggered by an event usually heightens its subsequent accessibility; it does not impair it.[52]

But when the idea of repression of memories of trauma came up, it was the repetition of the trauma and its intensity that was said to lead to repression. McNally pointed out that this belief was flawed:

If the frequency of sexual abuse increases the likelihood that one will be unaware of having been abused, then this would run counter to the ordinary effect of repetition on memory.[53]

Most psychological research shows that the longer the period of time that has passed, the worse is your memory; important events are remembered better than insignificant ones; and post-event information can distort or contaminate a memory.

Don Read and Stephen Lindsay, in a paper that tried to analyse the role of 'ordinary' forgetting in everyday life, where trauma was not involved, give a succinct definition of the distinction between forgetting and the much more serious condition of amnesia that is said to be at the root of repression due to trauma:

The term 'amnesia' connotes a pathological condition, created via special mechanisms distinct from those of ordinary forgetting. The construal of nonremembering of childhood trauma as a pathological condition carries with it the implication that nonremembering of trauma is a condition in need of treatment. Relatedly, some accounts of traumatic amnesia propose that the same special mechanism that impairs conscious recall of traumatic experiences also preserves hidden memories of the trauma in a preternaturally vivid and veridical form and that such hidden memories have maleficent effects unless and until they are recovered and integrated into conscious memory. Widespread acceptance of the ideas that amnesia is a common consequence of childhood sexual abuse, that hidden memories can be detected on the basis of symptoms, and that recovering such memories is an important aim or by-product of psychological intervention may lead clinicians inappropriately to foster memory recovery.[54]

The mechanisms that are said to produce repressed memories are very different from anything that has been discovered over the years about the normal processes of memory, and many researchers feel that the evidence produced so far is not strong enough to support the idea.

But this doesn't stop them looking for that evidence. Jonathan Schooler, a psychologist at the University of British Columbia in Vancouver, Canada, has done some interesting work on the idea that a motivation to forget something can lead to it not being remembered as effectively. He was intrigued by the idea of memory repression and didn't want to dismiss it out of hand, but, as he told me, he tried to think of a way in which it might come about without introducing entirely new psychological concepts:

I think one of the things that unnerves psychologists is the notion that there could be some kind of motivational system that is operating behind the scenes and is determining what is going to be made available to us and what is not, that we don't know about. This sort of 'intelligent unconscious' makes people rather nervous and I think this is one of the reasons that makes people wary about the notion of repression.

Schooler devised an experimental method based on a technique used to study people's personal goals and how to achieve them. It involves flashing the word 'smart' or 'intelligent' in front of experimental subjects for such a short interval that the person is not conscious of having read the word. The results showed that when the subjects are subsequently asked to solve a tricky problem, such as an anagram, those who were flashed 'smart' did better than those who were flashed a neutral or negative word. Similarly, if the subliminal word was 'elderly' or 'old', people moved more slowly after the experiment.

'The notion is that we have these unconscious goals that can be activated,' Schooler told me. 'You have to have at least a concept of the goal, and something as simple as "smart" could be the equivalent of a goal'.

Schooler adapted this technique by flashing the word 'forget' or 'remember' as the subject was shown a list of words for a few seconds. The results were interesting, but also illustrated the pitfalls of drawing conclusions too soon. Sure enough, the people in the 'forget' group seemed to achieve the goal of forgetting more often than the people in the 'remember' group. The effect wasn't huge, but it was statistically significant, and suggested that it was possible to 'learn to forget' or repress as some might put it.

The first experiment had used just two groups, 'forgetters' and 'rememberers'. So the question arose: were the forgetters really forgetting more or the rememberers remembering better? Schooler and his colleagues did a similar experiment, but with a third group added who were exposed subliminally to neutral words.

'What we found,' Schooler said, 'is that it was not that the "forget" instruction was leading to forgetting but the "remember" instruction was leading to remembering'.

Interesting as these results are, once again they involve people, usually psychology students, performing in an artificial situation, and inferences being made from the results about the natural processes of memory. But Schooler and some colleagues carried out another research project, this time investigating people who had been abused in the past and then claimed to have forgotten, or repressed, the memory.

They studied four such cases: a man of thirty-nine who had been abused as a child by his priest; a forty-year-old woman who had been raped at knifepoint at the age of sixteen; a fifty-one-year-old woman who had been molested at the age of nine; and a forty-one-year-old woman who had been raped at the age of twenty-two.

Each of these people claimed to have recovered a memory suddenly for the traumatic event after many years of being entirely unaware that it had happened. Schooler pointed out that three conditions needed to apply for these to be genuine examples of memory repression. First, there must be corroboration that the event happened. Second, there must be evidence that it had been forgotten. And, third, there must be evidence that it had been eventually remembered. These conditions seemed to apply with all four cases, but there was a surprise discovery. Three of the four turned out to have spoken of the trauma to someone else during the period where they were convinced that they had no memory. Like much analysis based on personal recall of events many years ago, actual hard proof of each stage in the argument—the corroboration of the initial event, the repression, and the remembering—was not watertight, but Schooler believed there was a genuine process which deceived people into thinking they had no memory of something over a long period, and he called this the 'forgot it all along' effect.

The papers reporting this research had quite an impact on some memory researchers, and confirmed the complexity of the interaction of memory and perception and the memory of remembering, and the dangers of saying that a memory had been entirely repressed. Richard McNally remembers hearing about this new work:

The example that I recall most vividly from Jonathan's work was a woman who had had experienced child sexual abuse, and then had this 'Oh my God!' effect of recovered memory and said 'I haven't thought about this in many, many years', so the husband says 'Wait a minute, you told me about this several years ago.' And so she says 'I did?' So here you've got a case of someone forgetting a previous recall, so it's tricky to document these things and it's sort of adventitious when it happens and very impressive.

Much of the laboratory research that is producing new insights into how memory works was triggered by cases like the George Franklin trial and its reliance on the 'recovered' memory of witnessing the murder. That was certainly what gave Elizabeth Loftus the impetus to change the direction of her research and look more closely at what was actually going on when people said that they were remembering something that happened to them in childhood which they had not remembered for many years.

Over the last twenty years, as a result of a series of pioneering experiments, Loftus has produced an impressive range of evidence showing how fallible childhood memories can be. A whole generation of psychology students, first at the University of Washington and now at the University of California at Irvine, have had their memories tweaked and distorted by the ingenious manipulations of Loftus and her graduate students.

Elizabeth Loftus testified in the Franklin case that there was no evidence that memories could be repressed in the dramatic and irretrievable way that Eileen Franklin-Lipsker claimed. But this was discounted by the jury, who believed another expert witness, Lenore Terr, who said in evidence:

...the return of a repressed memory is often like the bursting of an abscess: initially a huge bolus of memory shoots forth, and then, just as an abscess will continue to drain fluid (if treated correctly), the memory will release more details to the conscious mind. And as an abscess may never be fully cleaned out, so all the details of a repressed memory may never surface. In other words, blank spots in the memory are to be expected.[55]

The issue of expert evidence in criminal trials has become a battleground in its own right in the Memory Wars. At the Franklin trial, Terr went on to state in evidence a number of other scientific-sounding principles which struck Loftus as just plain false, and offered no scientific evidence to support them. For example, Terr described how to tell the difference between a false and a true memory:

First, the person's symptoms. Although the person may have repressed the memory, a part of them remembers the event and causes them to repeat it again and again through such behaviors as self-mutilation or suicide attempts, which are a result of the repressed rage and feelings of hopelessness. You should also see a handful of convincing symptoms, such as nightmares, intense fears, a lack of belief in your own future, or a feeling of emotional numbness. These children prime themselves to be escape artists, to hide, for example, in closets.

Second, a true memory should be rich in detail. A false memory is a child describing her psychiatrist dressing up in leather 'G-straps' and laying [sic] down on the floor and 'having sex' with her. A true memory has details about what it felt like, what it sounded like, a positional sense.

Lastly, a true memory should be told with the appropriate emotion; it should be accompanied by some weeping or some tightness of the body or some kind of a sign that emotionally that you're wrought up, that you're experiencing something.[56]

Terr was giving evidence in 1990, at a time when detailed research into what distinguishes true from false memory was in its infancy. But what is now clear, as I'll show later in the book, is that each of Terr's three defining characteristics of a true memory is wrong. There is no evidence that self-mutilation, suicide attempts, nightmares, intense fears, or 'a lack of belief in your own future' are specific signs of early child abuse; richness of detail turns out to be more characteristic of false memories than true ones; and people can experience intense emotional reactions when 'recalling' traumatic events which never happened.

'Expert witnesses' in trials are expected to base their testimony on evidence, and the kind of claims made by Terr, which are typical of what was, and still is, said by people who believe in the reality of repressed memories, sound like evidence-based statements. When Terr said in court that if a 'recovered' memory is accompanied by emotion it is a true memory, the jury was entitled to believe that the statement was based on rigorous analysis or experimentation. In this case, as Harvard psychologist Susan Clancy was later to show, while true memories may be accompanied by emotion, memories that are accompanied by emotion are not necessarily true.

The research of Elizabeth Loftus and other similarly sceptical psychologists in the last fifteen years has given the lie to many of the beliefs about childhood memory for abuse that were at the heart of a series of court cases in the 1990s.

Interestingly, as the queen of sceptics about repressed memories, Loftus has experienced what it is like to recover a memory suddenly, in connection with her mother's death by drowning when Loftus was fourteen. On her forty-fourth birthday, at a family gathering, she was told by an uncle that she had been the one to discover her mother's dead body. Until then, she had remembered little about the death itself, but she began to unearth clear memories of the events, just as had happened to George Franklin's grown-up daughter.

It was a few days later that Loftus learnt from her brother that her uncle had made a mistake, and that it had been Loftus's aunt who had found the body, not Loftus. So those few days of 'recovered' memories were utterly false. 'My own experiment had inadvertently been performed on me!' Loftus wrote. 'I was left with a sense of wonder at the inherent credulity of even my skeptical mind'.

In the Franklin case, in addition to testifying that she had recovered memories of her father killing her schoolfriend, Eileen Franklin-Lipsker also testified to serious episodes of violence and sexual assault at the hands of her father, which she had never mentioned to anyone in the previous twenty years. Loftus was puzzled by this. She could think of no corroborated example of where a child had suffered such traumas over a long period and entirely repressed the memory. In fact, it seemed to her impossible. At that stage, her memory research on suggestion had all pointed in the other direction—that it was very easy to produce memories of sights and sounds that had never occurred:

'I began showing people films of traffic accidents,' she said. 'In one early study, we found that questions such as "How fast were the cars going when they smashed into each other?" led to higher estimates of speed than a more neutral question that used the verb "hit". Moreover, the "smashed" question led more people to falsely claim later that they had seen broken glass when there was none. These studies were amongst the first to show that leading questions could contaminate or distort a witness's memory, but many others would follow'.[57]

Eileen Franklin's testimony in her father's trial involved a far more serious incident than watching film of a traffic accident, but, knowing the fallibility of memory, Loftus was suspicious:

As a scholar of memory, I wondered 'Did Eileen really repress all that brutalization and then remember?' In my role as a consultant and expert on the case, I investigated the evidence for such massive repression, and found there was virtually no credible scientific support for it. This was quite a surprise to me. The notion that traumatic events could be banished from consciousness appeared in novels in the mid 1800s, but Freud's popularizing of the concept of repression gave it legs and a place in psychological thinking. It began to trouble me deeply that someone could be convicted of murder based on virtually nothing other than a claim of repression and de-repression. Also I had another theory about where all the details of Eileen's 'memory' could have come from. Those details that gave her story such apparent credibility were widely reported in the media and were thus available to anyone who read the papers, watched television, or listened to others talk about the murder. Perhaps she incorporated these details into her memory, creating a belief that she had witnessed them personally.[58]

As it turned out, Loftus's instincts were right. In 1997, after several failed appeals at state level, Franklin's case was sent to the federal court, where it was argued that while the jury had relied on the statement in court that the content of Eileen's 'memory' could not have been gathered from the media, there were details in her story that were shown to be in the public domain. Since there was no other corroboration of the allegations than Eileen's testimony, and it was now unsafe to rely on that, Franklin was released. It had also come out during the appeals that the memory was not as 'repressed' as had been claimed. Five years before Franklin's arrest, Eileen had said to her husband, 'I think my father killed her'. Another example, perhaps, of Jonathan Schooler's 'forgot it all along' effect.

It is often difficult to trace why trends in society suddenly arise, seemingly overnight. At the moment, in British society, paedophilia and terrorism are part of the *zeitgeist*. Prosecutions occur on a regular basis of people who access images of child sexual abuse—or information about terrorist groups—on the internet; laws forbid people to take photos of their children in the nude or to allow them to play in the nude on the beach—or to own or distribute literature relating to terrorist groups; people are harassed because they are suspected of being paedophiles—or sympathetic to the aims of terrorists. None of these activities was so assiduously restricted or pursued ten years ago, but in one case at least, the focus on terrorism, we can put it down to the specific events of 11 September 2001. With paedophilia there is no such dramatic precipitating event—it just seems to be 'something in the air'. Undoubtedly, the easy availability of materials via the internet has produced more instances of people accessing such images, and created a climate in which there is concern that these people will go on and commit sexual crimes against children.

In the case of repressed memories of childhood abuse, there was arguably a '9/11' that has led to an explosion of such claims. It is a book called *The Courage to Heal*, by Ellen Bass and Laura Davis, which was published in 1988 and is still sold and promoted. Its subtitle is *A Guide for Women Survivors of Child Sexual Abuse*, and it offers advice and practical help for women who were sexually abused as children. While the exercises and procedures it suggests might be helpful for genuine victims of abuse, there is much in the book that is aimed at women who might never have considered themselves victims until reading *The Courage to Heal*. A more appropriate, if wordier, subtitle in my view would be *A Book Which Makes Women with a Wide and Diffuse Range of Mental Symptoms Believe that they were Almost Certainly Sexually Abused as Children*.

Mark Pendergrast is a writer who was accused by his own children of child abuse, and he spent several years talking to psychologists, psychotherapists, victims of child abuse, and parents accused of abuse by their children. He has no doubt about the malign effect of *The Courage to Heal*. He compares the ongoing search for episodes of child abuse and for abusers based on recovered memories with the witch-hunts of fifteenth- and sixteenth-century Europe:

The zealous clerics and judges who ferreted out these evil witches had help from numerous manuals which described the symptoms of witchcraft in great detail. The first and most famous, the *Malleus Maleficarum*, or 'Hammer of Witches'... was published in 1486. It is a remarkable document which, like *The Courage to Heal*, offers an internally logical and quite convincing way to identify the root cause of the problem. In this case, however, it was witchcraft rather than repressed memories of sexual abuse that wreaked havoc in people's lives.... Compared to *The Courage to Heal*, however, the *Malleus* was in some ways a moderate, well-reasoned document.[59]

It's interesting to note that during the medieval witch-hunts, hysteria and social pressure led to accused women actually believing that they were witches, by acquiring false memories:

They had developed very coherent, detailed memories of the orgies in which they had taken part, the babies they had roasted and eaten. Young girls would describe in gory detail how they had been deflowered by the Devil, though examination proved them to be virgins.[60]

Richard McNally also believes that *The Courage to Heal* is to blame for encouraging people, usually women, to accept that they were sexually abused as children on the basis of a range of signs and symptoms that in fact many unabused people display:

In The *Courage to Heal*, Ellen Bass and Laura Davis informed readers that 'many adult survivors are unaware of the fact that they were abused,' owing to the mind's 'tremendous powers of repression.' Hidden abuse, they said, could be revealed in many ways, including feeling 'different from other people' and feeling a need to be 'perfect'. Bass and Davis reassured readers: 'If you don't remember your abuse, you are not alone. Many women don't have memories, and some never get memories. This doesn't mean they weren't abused.' For those who questioned the authenticity of their memories, they wrote: 'So far, no one we've talked to thought she might have been abused, and then later discovered she hadn't been. The progression always goes the other way, from suspicion to confirmation. If you think you were abused and your life shows the symptoms, then you were.'[61]

(As a matter of interest, as we will see later in this book, there are countless examples of women who thought they might have been abused, as a result of *The Courage to Heal*, and later discovered they hadn't been.)

Mark Pendergrast assembled an impressive number of testimonies from women who were set on the path of 'recovering' memories of being abused as children as a result of *The Courage to Heal*:

The Courage to Heal is the BEST book—very validating if you have no memories.[62]

My therapist told me to read *The Courage to Heal*. I opened the book to the first page, and three hours later, I looked up, sobbing. I was totally consumed by this book. I couldn't read enough, find out enough, couldn't let it go. Everything was leading me down this road. My therapists weren't necessarily saying 'Confront your parents,' but society and books and my need to be healthy were driving me. I was absolutely driven.[63]

I gobbled up The Courage to Heal—just read it, read it, read it, particularly the stories by Survivors at the back of the book. They were so awful. The phrase kept coming back, 'If your life shows the symptoms and you don't remember it, you were still abused.' I just lived with that phrase. And I bought Secret Survivors and Silently Seduced and a bunch of other recovery books, too. I read them all. By the time I went for my next appointment, I was an Incest Survivor.[64]

[Karen, a therapist] had groomed me for over four years to get to this point. In desperation to get well, I said I was willing to entertain the thought [of repressed memories]. Karen gave me a copy of The Courage to Heal, and soon after that, I succumbed completely and became a Survivor.[65]

The Courage to Heal was soon turned into a 'brand', with workshops and audio tapes and workbooks, all promoting the idea that symptoms that had previously been put down to unspecified psychological disorders were actually caused by childhood events that the sufferers had entirely forgotten. Mark Pendergrast quotes from an interview with a woman who attended one of Ellen Bass's seminars and as a result made a horrifying discovery:

I always knew that when an emotion twinged at whatever somebody was saying, it probably had something to do with me, and I should pay attention to that. But that day, there were no emotions to go on. So I thought, well, I'll make a note in the margin of my journal whenever I have a *physical* discomfort. I made one mark in the morning when I had a real brief little twinge, then another one in the early afternoon. Late in the day, it happened again. And that third time, [Bass] had just done a ten minute talk about ritual abuse. Without gory details, she presented just the minimal facts, but just the idea that babies were sacrificed was so horrible the room was in shock.

For this attender at the seminar, her twinges eventually left her in no doubt that she herself had been the victim of ritual abuse:

One of the worst memories I have is of being buried alive, and the sacrifice that preceded it. The basic message of the sacrifice was proving that love has no power at all. I was about five. They murdered a baby, then they cut off his mother's arm. I don't know what happened after that, because they buried me alive, put me in a coffin. When they dug me up, I was completely blue and stiff, yet I wasn't quite dead. Getting that particular memory up has been one of the hardest parts of my recovery. It's taken me more than a year to get the pieces I have.[66]

For Bass and Davis, nothing seems too implausible if it emerges—or can be made to emerge—from the imagination of disturbed people. Stories of murdered babies abound, and the fact that these little corpses are never found does not seem to dent the tenacity of Bass and Davis and the therapists of their school. Asked why there is no evidence of such extreme abuse, therapists give the sort of reply one psychiatrist gave to Mark Pendergrast:

The critics ask for evidence, for the bodies of murdered babies. One wonders why there isn't any evidence. I am not an expert on such matters. It is the social workers and church counselors who have the most experience with ritual abuse survivors. They claim that the lack of evidence can be explained by cult members in high places very cleverly hiding it. They believe that there are doctors, lawyers, politicians, and mortuary attendants involved.[67]

At the root of the 'memory recovery' process promoted by *The Courage to Heal* is the hypothesis that if someone suffers extreme trauma when young, the mind can repress that trauma so deeply that special techniques are required to 'recover' the memories and bring the abuse to light.

In the terms in which the more informed memory repression supporters state their theory, it is a testable hypothesis. Not a true one, necessarily, but it can certainly be a starting point for scientific research which will establish whether it is valid or not. It is more scientific than, say, the statement that, when children are small, incubi appear in the middle of the night having assumed the form of their fathers and abuse them, sprinkling fairy dust over them as they leave the room, so that they forget everything. No experiment or research project could disprove that theory and so it is unscientific.

While there was considerable scepticism about the concept of re-pressed memory, it was not immediately obvious why someone would lie about having sudden vivid memories of a childhood event. On the other hand, there were very plausible arguments for why it might be useful for the mind to suppress a memory of an extremely traumatic event, and why it might nevertheless make its presence felt through bodily symptoms, dreams, even suicidal behaviours for no apparent reason. The problem was, at the time, these were all suppositions. There was no evidence. By definition, if these things happen to some-one as a result of a repressed memory, there is no way of knowing of the existence of that memory. It's pure hypothesis.

But people *need* explanations for their problems, and when a young woman experiences unexplained symptoms—eating disorders, depres-sion, anxiety, nightmares, sexual problems—and she seeks therapy, it is very reassuring to be given a cause, even if you have no recollection of any abusive events. There is even the feeling of having come in from the cold, of being part of a larger community of sufferers.

'There is an identity in being a committed survivor of sexual abuse,' Bass and Davis wrote in *The Courage to Heal*. 'It can be hard to give up'.[68] One 'survivor' confirmed this to Pendergrast:

Eventually, I accepted that my father had abused me. I became very with-drawn. I still saw my parents, but very infrequently, I didn't say anything to them, but I started to reveal these terrible awful secrets to others. The more people I told who accepted it, the more I thought it must be true, I told all my friends, even very new friends, I was getting my strokes and my attention from these people. You know, 'Oh, you poor thing'.[69]

'This all gave me a key to everything,' said another young woman. 'It explained why I lost my job: I had transferred my feelings towards my father to my new boss. He was trying to control me and being abusive, just like my father had been'.[70]

Loftus, McNally, and many other psychologists could not accept that when psychologically disturbed people came to 'recover' memories for traumatic events that occurred in their childhood those memories were necessarily accurate, particularly if those memories had been entirely absent for years or decades until 'recovered' in psychotherapy or as a result of reading *The Courage to Heal*. Surely, they thought, this is a matter that science could help to settle?

But before the science, some homework. Read the following list of words once carefully so that you think you would recognize them if they were embedded in a list of different words.

bed, rest, awake, tired, dream, wake, snooze, blanket, doze, slumber, snore, nap.

Now read on.

6

· · · · · · · ·

Playing False

John Ridley Stroop was an unlikely originator for an experiment which was to appear in every psychology textbook and become known as the Stroop Effect. His PhD thesis in 1935 was one of his few written contributions to psychology, before he embarked on an entirely different course which led to what he considered a far more important work, a Christian trilogy called *God's Plan and Me*. In later life, when he was known as Brother Stroop, he concentrated on teaching and promoting the Bible in his native Tennessee. But the technique he developed turned out to be very useful to Richard McNally and his colleagues when they investigated whether trauma in childhood leads to memories of the events being repressed.

Stroop's PhD was based on an odd and easily demonstrable effect which occurs when subjects are presented with a word which is a colour name but printed in a different colour—the word GREEN, say, printed in red ink. Asked to name the colour of the ink, people consistently take longer than with a control word, a neutral word like 'blank' printed in red ink. And however hard they practise, the subjects can't improve on their performance. This simple task has been used to study attention and the way the brain processes language and meaning, and more than a thousand studies have since been published using this method. It can be adapted to test other semantic categories: for example, the words 'two two two' are presented and the subject asked to say how many words there are.

When Richard McNally came to look at the evidence that traumatic events in childhood can be entirely forgotten, there was really only one piece of research that suggested this, by Lenore Terr, who gave evidence at the Franklin trial. She had studied twenty children who had been traumatized before the age of five, some on a single occasion, others repeatedly. She was comparing two types of trauma, repeated and single-incident. Her conclusion was that forgetting did occur and that repeated traumatic events are more likely to be forgotten than a single traumatic event. But, as Richard McNally pointed out:

The children in the single-trauma group had been much older than those in the repeated-trauma group when the adverse events occurred. Indeed, the 3 children in the repeated-trauma group who had no verbal memories had been only 6 months, 24 months and 28 months old when their trauma ended. No special mechanism is needed to explain their lack of narrative memory for events occurring prior to the offset of childhood amnesia.[71]

In any case, on looking at other investigations of trauma sufferers Terr's data seemed rather out on a limb. A study of survivors of motor accidents who were suffering from acute stress disorder showed a marked increase in trauma-related thoughts when they attempted not to think about the accidents. Similarly, in a survey of rape victims, when they attempted to suppress trauma-related thoughts in the laboratory the thoughts became even more frequent.

Surveys of survivors of the Nazi holocaust also show the difficulties of forgetting the traumatic experience. Wagenaar and Groeneweg studied memory for trauma among former inmates of Camp Erika, a Nazi concentration camp, and found that they showed 'a remarkable degree of remembering' of their dreadful experiences. 'There is no doubt,' they said, 'that almost all witnesses remember Camp Erika in great detail, even after 40 years'.[72]

One psychologist, Richard Gardner, has written:

I cannot imagine such a person saying, 'Concentration camp? What concentration camp? I have absolutely no memory of the experience,' or 'I had obliterated entirely any memories of having been in concentration camp,' or 'The whole experience was totally repressed. However, in the skillful hands of my therapist I gradually recovered memories of the cattle cars, firing squads, rape of women, gas chambers, and the crematoria'.[73]

For Gardner, to accept that total repression of childhood memories of trauma and abuse is possible would require acceptance of 'A new set of psychological principles, principles totally at variance with everything we know about childhood trauma, our capacity for repressing it, and its residual manifestations in adult life'.[74]

Elizabeth Loftus, too, is convinced that repression is a myth:

This idea of repression, that takes years of brutalisation and banishes it into the unconscious, is supposed to be a process that's beyond ordinary forgetting and remembering. Nobody doubts that you can 'not think' about something for a while and be reminded of it, or that retrieval cues can trigger your thinking about something that you haven't thought about for some time. But what you have to believe in order to believe some of these repressed memory cases is something so extreme that ordinary forgetting and remembering can't explain it.

Richard McNally has spent the last ten years investigating whether the memory of trauma can be so repressed that it is impossible to retrieve it in the normal course of events. On a damp September afternoon, I went to visit him in his twelfth-floor Harvard office with a view towards downtown Boston, its skyscrapers almost obscured by low clouds and fine rain. Hastily finishing a sandwich, he bounded towards me and ushered me towards a squishy leather sofa into which I sank and almost lost my balance as I did so. A man in his forties, he wore a red T-shirt and spoke in a strong voice, punctuated by laughter as he described some particularly outrageous piece of pseudo-science, or told of attacks on his work that had come from people who preferred their own views of repressed memories to the often disappointing and sometimes dull data that have emerged from the careful experiments of McNally and his colleagues.

There are many accounts from abuse survivors whose problem is that they can't forget, not that they can't remember. It might be wondered why McNally and other psychologists in the 1990s embarked on a search for evidence for a process which seems, on the face of it, implausible. But there were so many people promoting the idea of repressed memories at that time, some of them more scientifically qualified than the authors of *The Courage to Heal*, and people were actually being sent to prison on the basis of the ideas, that McNally felt the theory had to be investigated:

I was sceptical for a number of reasons. First of all, I was trained in the cognitive behavioural therapy tradition. I'm not a Freudian, I'm not psycho-analytically inclined. Freud once said there was supposedly one race of people whose mind is so defective that psychoanalysis can't touch it, they're immune to it, and that's people of Irish ancestry, they're hopeless. It's Irish American, in my case, so that may be my problem, my Irish genes.

To investigate how impressions of traumatic experiences are dealt with in the brain, and how easily they can be repressed, Brother Stroop's innovation came in handy for McNally and his colleagues, in the following way:

People who have experienced severe trauma, involving sexual or physical abuse, or witnessing or being subject to violence, sometimes suffer from what is called Post-Traumatic Stress Disorder, or PTSD. The symptoms include flashbacks, numbing of feelings, nightmares, clinical depression, and anxiety. In a study using the Stroop method to investigate thoughts of trauma in Vietnam veterans who suffered from PTSD, the subjects were presented with four categories of word printed in different coloured inks, and asked to name the colours, not the words. There were trauma-associated words, like *firefight*, negative words, like *faeces*, positive words, like *friendship*, and neutral words, like *concrete*. The PTSD survivors took significantly longer to name the colours of the words associated with trauma than any of the other three categories. For the researchers, this indicated that far from being repressed, ideas associated with the trauma the veterans had suffered were continually intruding into their thoughts and getting in the way of normal cognition.

The horrors of battle and the PTSD that follows are different from sexual abuse and its after-effect, not necessarily worse, but different. McNally and his colleagues decided to assemble three groups of people in order to study the effects of childhood sexual abuse specifically. The groups were: women with a history of sexual abuse and PTSD; women with a history of sexual abuse but no PTSD; and women with neither abuse histories nor PTSD.

They used another experimental method which used a range of emotive or non-emotive words as stimuli. There were trauma-related words (such as *incest*), positive words (*celebrate*), and neutral words (*mailbox*), and they appeared one at a time on a computer screen. The subjects were asked either to remember or to forget each word, and later they were asked to write down as many words as they could remember, ignoring the original instructions. Far from repressing the trauma-related words, the subjects in both groups remembered more of them than the pleasant or neutral words. Other psychologists using similar techniques with related groups found a similar tendency to remember negative words rather than forget them more readily. McNally summarized the results:

The preponderance of laboratory research confirms reports of intrusive recognition in trauma survivors with PTSD. Emotional Stroop studies show that survivors with PTSD have difficulty suppressing the meanings of trauma-related words. Moreover, despite their strong motivation to forget traumatic material—or perhaps *because* of this motivation—survivors with PTSD have trouble forgetting such material. These laboratory studies directly contradict the hypothesis that survivors are especially capable of forgetting trauma. Judith Herman, a PTSD researcher, once wrote: 'The ordinary response to atrocities is to banish them from consciousness.' But if survivors have difficulty banishing mere trauma-related words from consciousness, how much more difficult must it be to banish atrocities? *Attempts* to banish traumatic memories from consciousness must not be confused with *success* at doing so.[75]

There's another factor which complicates the picture. If people lack memory for some important aspect of an experience, or even the whole experience, how do we know they ever took in the experience of the event in the first place? Psychologists talk about 'encoding', as part of the linked processes of perceiving and remembering. There's a vivid demonstration of how we sometimes fail to encode quite important parts of an experience, on a video clip produced by the Visual Cognition Lab at the University of Illinois. The instructions that go with the clip are: 'When viewing the video, try to count the total number of times that the people wearing white pass the basketball. Do not count the passes made by the people wearing black'.

The web address to view this clip is:

http://viscog.beckman.uiuc.edu/grafs/demos/15.html

The significance of this exercise, if you view the video, is explained in this endnote:[76] (If you read the endnote before watching the video, the exercise will be pointless.)

One believer in the reality of repressed memory is Jennifer Freyd at the University of Oregon, and she and McNally spar from time to time in the scientific journals. Freyd believes that there *are* extreme situations in which children who have been abused grow up with no memory of the abuse. As a good scientist she has tried to come up with an explanation for a process which might, as I've suggested above, seem illogical and not evolutionarily helpful.

Freyd has a theory which she calls 'betrayal trauma'. She believes that if a close relative or friend abuses a child, a conflict is set up between the repulsion or avoidance reaction that might be induced in the victim and the need of the victim for the caring provided by the abuser.

'Victims of abuse may remain unaware of the abuse,' Freyd says, 'not to reduce suffering, but rather to maintain an attachment with a figure vital to survival, development, and thriving'.

Freyd believes that memories can be repressed to such an extent that they are impossible to retrieve without extraordinary measures. In a situation where there is room for doubt about the validity of a theory, it helps if you can discover a plausible explanation. Freyd's theory takes an evolutionary approach and looks for why it might be adaptively useful to children to repress memories of abuse. She told me how she believes this sort of repression arises when a child, or indeed an adult, is dependent on a certain relationship, and where some negative event occurs that endangers the relationship:

When people detect that something is unfair—for example, they detect they've been swindled—if they're adults they'll usually react in ways that are not conducive to a friendly relationship, so they'll withdraw and say 'forget it,' or they confront and they say 'don't ever do that to me again', and they yell and they scream, or whatever. But those behaviours can be really problematic in a relationship where you're dependent on one other person and that other person has power over you. In the case of children it's not just confrontation that can be a problem, it's also withdrawal, because if a kid withdraws from a caregiver relationship they're in danger of not getting as much nurturing, and so withdrawal is not a good thing for a baby or a child to do. With betrayal trauma it's really dangerous for the child to stop behaving in ways that keep the relationship going, which would be the way an adult who has detected cheating would behave.

So, to make it possible to continue in the caring relationship, Freyd believes, we have evolved to repress the memory that might lead to conflict, the memory of trauma and abuse, so that the child can continue to believe in a loving, non-abusive carer.

One consequence of this theory is that memories of child abuse would be repressed more if the abuse is carried out by parents and relatives and those on whom the child depends, than by strangers or friends. Jennifer Freyd believes she has found evidence that this is the case; Richard McNally and others find her data weak and have produced other data which show no connection between the dependency relationship between a child and the abuser, and the forgetting or repression of memory for that abuse.

'Despite the evidence against betrayal trauma theory,' McNally points out, '—and the meagre evidence consistent with it—this theory remains popular among therapists who are convinced that many incest survivors cannot ever recall having been abused. The discrepancy between evidential support and popularity could not be more striking'.[77]

Jonathan Schooler, who finds the betrayal trauma theory a plausible one and has looked for evidence himself, nevertheless said to me:

I'm not aware of much compelling empirical evidence to support it. In the sample that we looked at we did not find any correlation between the relationship between the abuser and the victim, and whether or not the memory was characterized as having been forgotten.

The fact that some people might not remember episodes of traumatic abuse does not of course mean that traumatic abuse causes forgetting. It is necessary to compare remembering or forgetting of *non*traumatic events to see if there is a significant difference between the two groups.

Don Read and Stephen Lindsay, two Canadian psychologists, have done just that in a study of people's memories of a series of nontraumatic life events, including 'summer camp', an experience most North American middle-class children undergo on a regular basis. In an important paper published in 2000, Read and Lindsay summarize the main findings of research linking forgetting with trauma, and then they write: 'Implicit in the conclusions of such studies is the assumption that retrospective reports of partial or complete amnesia for comparably memorable but nontraumatic life events would be at or near zero'. In other words, people don't forget harmless incidents from their childhood.

In their experiment they asked about forty adults to rate their current memory for a series of childhood or adolescent events they had listed two years before as having experienced. The first finding was that a small but significant percentage had completely forgotten events that they said they had remembered two years beforehand, events such as music lessons, Boy Scout or Girl Guide activities, or summer camp. This showed that partial amnesia occurs for non-traumatic events in some people. They were also asked to estimate the percentage of recall they now had for partially remembered events, compared with what they would have remembered the day after the event. The overall average of such estimates was 15 per cent.

In the next phase of the experiment, lasting four weeks, the subjects were divided into three groups. One group, 'Reminiscence', was encouraged to think about the partially remembered events as much as possible, trying to remember more elements than they had at the beginning of the experiment; a second group, 'Enhanced', was encouraged to do this, and, in addition, to seek other reminders of the experiences, by visiting the site, talking to relatives, and looking at family photos; and the third group, 'Control', did neither of these.

The results showed that activities like reminiscing about forgotten aspects of events led to a more detailed memory for the event, and actively seeking out reminders increased the amount remembered even more. At the beginning of the experiment 15 per cent had claimed that they had partial amnesia for events they had experienced; after the period of 'Reminiscence', the figure went up to 35 per cent, and in the 'Enhanced' group to 70 per cent. Lindsay and Read summarized their conclusions:

Our point is that reports of prior periods of partial or complete 'amnesia' may be common for nontraumatic events and that this must be taken into account when interpreting data on reports regarding memory for traumatic events. For example, Gold et al. noted that, following therapy, their respondents spontaneously reported their pretherapeutic memories to be much more fragmentary than they had originally assumed. Our results indicate that such reports are likely to follow any period of sustained recollective activity about life experiences, rather than being peculiar to memories of traumatic events.... Overall, these findings should not be surprising, but they have obvious implications for the interpretation of studies in which participants who recently received trauma memory-oriented therapy, or who engaged in self-help memory exercises, or who have just completed a lengthy and detailed interview regarding their sexual history are asked about prior periods of less memory for childhood trauma.[78]

Psychology is sometimes seen as a science that produces less reliable results than, say, physics or chemistry or astronomy, in the sense that the subjects of psychology experiments—human beings and their central nervous systems—are far more complex than single atomic particles or even than a star cluster or spiral galaxy. You are also rarely able to get near the specific physiological sites of psychological concepts—memory, depression, love—and are reduced to treating the person as a black box responding to inputs with outputs. The job of the experimental psychologist is then to infer what has intervened between the input and the output and create more black boxes which through connections with each other make up the processes that are responsible for human psychology.

When psychologist Matthew Hugh Erdelyi wrote that 'Repression has become an empirical fact',[79] he was taken to task by other psychologists who asked:

Is it an empirical fact in the same way that, say, gravity is an empirical fact? Well, no. We can demonstrate the effects of gravity simply, quickly, and whenever we want; the same is not true of repression. Is repression a widely accepted theory, in the same way that, say, the theory of evolution is? Well, no. Converging evidence from biology, biochemistry, and anthropology supports evolution. This is not true of repression. Is repression an idea that sounds interesting but turns out to be something that nobody can find any evidence for—like, say, cold fusion? Yes. That's more like what repression is: cold fusion. Repression has become the clinician's cold fusion, at once obvious and problematic.[80]

Throughout his work on memory Richard McNally aims for the same kind of rigour that accompanies the experiments of physicists, molecular biologists, or X-ray crystallographers. This may seem unsurprising, but there are plenty of psychologists who feel that because of the difficulties of getting unambiguous answers in psychology, the standards of conventional science should not be applied to psychology. Jennifer Freyd herself has written a paper called *Science in the Memory Debate* in which she argues that 'memory science is important in this debate but it is not the only domain of knowledge that needs to be considered'. This is dangerously near to the stance of people who are supporters of psychic phenomena and who, when science fails to demonstrate the existence of such phenomena, say that there are other ways of gaining knowledge. Freyd calls for 'philosophical analysis of epistemological complexity, sociology, history, analyses of the interplay between power and knowledge',[81] all to be taken into account, since memory science falls short of demonstrating the existence of repressed memories. (This last point is an inference from her paper, but if memory science *did* prove the existence of memory repression why would we need recourse to such things as 'philosophical analysis of epistemological complexity'?)

This is not to say that science in the psychology lab can settle the matter conclusively. McNally himself admits there are limits:

First of all, obviously we cannot manipulate memories of abuse in any real way, so there's ethical barriers to answering certain kinds of questions. What we *have* done is experiments on directed forgetting—in other words: are people who report a history of forgetting child sexual abuse, or who harbour repressed memories, possessed of a well-honed skill to disengage attention from words related to trauma and block them out and have a hard time remembering them later? We've done a number of experiments attacking this from a number of different ways and basically the story is that it's very tough for people to forget trauma-related materials, including those with a history of child sexual abuse. Now there's some obvious limitations here—we're talking about words, mere pale proxies for autobiographical memories of sexual abuse. You're not sure which way this would go. In other words you might say 'Well heck, if these were real memories then the motivation to forget would be much greater and therefore we'd see repression'. On the other hand you could say 'if you can't even block out memories of a word related to trauma, how are you going to do this for a sensory-rich autobiographical memory'. So you're not quite sure how the limitation goes, you might be underestimating it or overestimating it.

Whatever the limitations, McNally is genuinely trying to answer the question—can memories of childhood trauma be repressed? He is prepared for a result either way, even if, like Elizabeth Loftus, he thinks the idea is inherently implausible.

Although there might be ethical barriers to some aspects of the work, it is possible to carry out retrospective studies of survivors of abuse which undoubtedly took place, and analyse their memories for the events.

McNally's interest in repressed memories started with his study comparing people who had been sexually abused as children and developed PTSD, with other survivors of childhood sexual abuse who did not. He had found the subjects by advertising in a local newspaper for people who had been sexually abused as children, and as he started interviewing people who responded, he came across a surprising finding with some of the subjects:

What happened was that I got to the point in the psychiatric interview in which I was asking the potential subject who the perpetrator was, what had happened when the abuse occurred and so forth, to get an idea of what the events were and then to ask about the symptoms. In each case the person said to me 'I don't know'. I was a little bit taken aback by that response . . . 'Did they misread the ad?' is what I thought. And I asked 'how is it that you responded to the ad soliciting adult survivors of child sexual abuse and you have no memory of having been abused?' and in each case they would say something like 'Oh well, I eat too much and I make myself throw up', or 'my mood often fluctuates and I don't understand why', 'I get tense around my stepfather for no obvious explanation', 'I have nightmares that I can't explain', 'I have sexual problems, what else could it be?' 'I must have been sexually abused and I can't remember it.'

McNally found this finding so surprising that, in collaboration with a colleague, Susan Clancy, he decided to study people who claimed to have been abused as children but had no memory of the abuse. The research that he and his collaborators carried out was to provide data in the place of rumour and superstition, and show that people have all sorts of reasons for believing strongly in events from their past, many of which have nothing to do with whether the events happened or not.

'When a friend recently asked me about my happiest memory, I didn't have to think twice,' wrote Susan Clancy in a book about her research. 'I immediately recalled a day in Aspen. It was my day off, there were three feet of new powder on the slopes, and I'd gone skiing with my new boyfriend, a gorgeous Australian ski instructor. It was late afternoon. We had just skied off the back line, and we were on the roof deck of his condo in a hot tub. It had started to snow, and big, beautiful, fat flakes were melting in his golden blond hair'.

You can just see the scene, vividly described, accompanied by an aura of happiness, just the sort of memory anyone might have. Reconstructed, certainly, as we now know, but not necessarily wrong. Only in this case it was.

'When she heard this story, my friend laughed,' Clancy continued, 'and told me to pick another field to study. Why? Because she'd been there. She reminded me that I hadn't been having so much fun. I'd been wearing the wrong skis for powder and had spent most of the day falling down; the boyfriend had kept yelling, "Quality air time, dude!" every time he hit a mogul; and I'd been coming down with the flu. I hadn't really wanted to be in the hot tub; I'd had to be at work at six; and my borrowed bathing suit had been saggy and kept filling up with bubbles. It hadn't even been snowing'.

Even Clancy, who knows a lot about memory, had distorted her own recollections in the direction of what she had wanted to happen rather than what actually had happened.

'Snuggling with a ski instructor in a hot tub after a terrific powder day is the kind of memory I would like to have,' she wrote. 'It's consistent with my image of what I used to be (or what I wish I had been), and with what I would expect to have as a best memory. A moment of perfection in my life history was shattered by some inconvenient facts'.[82]

This recollection has many aspects of a 'false memory', by which I mean more than memories that are merely incomplete, inaccurate, or imprecise. When I try to remember my last school prize-giving, I think my prize was a biography of Evelyn Waugh, but it might have been another book called *Mass, Length and Time*; I remember first visiting Niagara Falls in January but it might have been December; I remember that the first Elvis Presley song I owned was *All Shook Up* but what if it was *Blue Suede Shoes*? To me these are not really 'false memories'. False memories are for things that didn't happen at all, and may have been created or planted in some way. As Susan Clancy was reminded, it wasn't snowing, so the flakes weren't melting in the Australian's hair, and he wasn't gorgeous but irritating.

Here's a childhood memory, recounted in a book review in the *New York Review of Books* by the distinguished, and generally accurate, scientist, Freeman Dyson:

When I was a boy in England long ago, people who traveled on trains with dogs had to pay for a dog ticket. The question arose whether I needed to buy a dog ticket when I was traveling with a tortoise. The conductor on the train gave me the answer: 'Cats is dogs and rabbits is dogs but tortoises is insects and travel free according.'[83]

It's the sort of thing which could happen, and if it did, might be remembered reasonably accurately. But the British psychologist Nicholas Humphrey wrote to the *New York Review of Books* explaining why it was almost certain that this was a false memory:

Freeman Dyson recommends that we should try to understand other people from the inside. But it does seem to be carrying this process of identification a bit far for Dyson to attribute to his own childhood experience the encounter with a train conductor concerning the status of a tortoise.... The very same encounter appeared as a cartoon in *Punch* in 1869. The caption of the cartoon reads: 'Railway Porter (to Old Lady travelling with a Menagerie of Pets). 'STATION MASTER SAY, MUM, AS CATS IS "DOGS," AND RABBITS IS "DOGS," AND SO'S PARROTS; BUT THIS 'ERE "TORTIS" IS A INSECT, SO THERE AIN'T NO CHARGE FOR IT!' [*Punch*, 1869, Vol. 57, p. 96]

Dyson replied: '[One explanation] is that I heard of the conversation recorded in the *Punch* cartoon and transformed it over the years into a memory. This would not be the first time that I remembered something that never happened. Memories of childhood recollected in old age are notoriously unreliable'.

And his other explanation? Well, it seems that false memories can be so vivid that the rememberer is sometimes reluctant to accept the falseness. 'The second possible explanation,' Dyson wrote valiantly— and implausibly, 'is that the memory is accurate. In that case the conductor on the train knew the cartoon and said what he was supposed to say according to the script'.

If we're interested in how human memory deceives us, such anecdotes as Susan Clancy's or Freeman Dyson's are not very useful. There's no experimental way of discovering the conditions at the time which might have led to the false memory. But in the 1990s two psychologists, Henry Roediger and Kathleen McDermott, wrote an article in which they described a technique that could be used to induce false memories in the laboratory. It so happened that this technique had been devised more than thirty-five years beforehand, by James Deese, another psychologist, who was still alive and working at the University of Virginia. Maybe Deese was too shy and retiring to promote his technique, but after Roediger and McDermott described the method it turned out to be very useful in a number of ways, and became known as the Deese-Roediger-McDermott procedure or DRM.

'Deese published it first in the 1950s and was roundly ignored,' Richard McNally told me, 'until Roediger and McDermott stumbled across it again in the late 80s, maybe possibly 1990, and it all took off again. The first paper was '56 or something and people said "Oh, yes" but not *very* interesting. The false memory idea wasn't really fashionable at the time. It just seemed like an odd finding and was filed away'. At least D's initial was put first.

The DRM procedure is deceptively simple, which may be why it is frequently used in psychology classes to demonstrate the production of false memories. Think back to the list of words I put at the end of Chapter 5, but don't look at them again. Which of *these* words was on that list?:

drowsy, chair, bed, awake, rest, wardrobe, sleep, tired, yawn

If you remember reading *sleep* on the first list, you have acquired a false memory. It wasn't there.

In the DRM procedure, lists of twelve words like the list at the end of Chapter 5 were shown to students. Each list of words had something in common that was expressed by a word not on the list, called a 'lure'. In this case, the lure was *sleep*.

The experimental subjects were then tested. Some of the tests assessed which words they could recall, and others which words they could recognize.

In a rather more rigorously controlled experiment than mine, subjects *recalled* 65 per cent of the words that had been on the lists presented to them, but they also recalled 40 per cent of the words that had not, the lures. They *recognized* 86 per cent of the study words but also recognized 84 per cent of the lures.

Since R&M first revived D's procedure, and took some of the credit, it has been used countless times to demonstrate the production of false memories and also to explore the conditions under which they occur and how they might be explained. For example, experiments have shown that false memories induced by the DRM method persist, in some cases longer than real memories. They can also be induced by using pictures instead of words, omitting a 'lure' picture whose theme underlies the sequence of images. There have even been attempts to look at brain activity, using PET scans, or positron emission tomography. Daniel Schacter and his colleagues have shown that when words that were in the original list are recognized there's greater activity in a certain area of the brain than when the 'lure' word is recognized. It seems that the brain could tell the difference between true and false even if the subject was not aware of it.

Various explanations have been suggested for the effect. Presented with a list of words connected with sleep, we may at some point *think* the word 'sleep' and that then gets mixed up with the other words we're also thinking of, even though for a different reason.

Even if the word 'sleep' itself doesn't get lodged in our memory at the time, since it's a generic descriptor of the words on the list it may come up when we think back to the list at the time of testing.

At first glance, the DRM procedure may seem an artificial and arid process. Real-life false memories, like Susan Clancy's and Freeman Dyson's, are not single-word mistakes. They seem to bear as much relation to the DRM procedure as *The War of the Ghosts* to Ebbinghaus's nonsense syllables. But there's one major difference. Ebbinghaus was trying to avoid any hint of meaning or association by using nonsense syllables. DRM produces an effect that is *based on* meaning, perhaps by activating traces of remembered items with similar semantic qualities.

Nevertheless, Jennifer Freyd and a colleague have criticized the overgeneralization of laboratory results like the DRM procedure to the world of childhood and trauma:

The experiments and data presented by Roediger and McDermott (1995) are clean, crisp, and compelling in and of themselves. Yet, the bold speculations about generalizations to the current controversy about recovered memories of abuse— speculations made, with the voice and context of powerful scientific authority— have the potential to be used in courtrooms around the country to help support the position of persons denying having committed a felony and often actively attacking those who say otherwise.... If this laboratory science were truly applicable to the issues in the way Roediger and McDermott suggest, then these consequences would arguably be a necessary cost in achieving an understanding of truth. But in this case the science does not support the generalization . . . We urge readers to exercise caution in generalizing from laboratory results, to highly political, emotionally charged, real-world controversies.[84]

In spite of these concerns, Jennifer Freyd's worry about generalizing laboratory results to a broader context didn't stop her using the Stroop colour-naming task, a laboratory technique which some might say is equally removed from 'highly political, emotionally charged, real-world controversies' to support her own 'betrayal trauma' theory.[85]

But such doubts about whether laboratory results are comparable with real-life situations were discredited by a series of experiments carried out by Elizabeth Loftus, which cast a new light on false memories that were far more complex than words on a computer screen.

Loftus's early research work, before she ever got sucked in to the repressed memory maelstrom at the Franklin trial, had provided some evidence that it was possible for people to come to believe firmly in things that hadn't happened. In addition to the earlier road accident movie experiment, Loftus had carried out another one involving a car going through a 'Stop' sign. In later discussion of the incident with the subject, the suggestion was planted that it was a 'Yield' or 'Give Way' sign, and a significant number of the subjects incorporated that into their later memories, as Loftus described:

Of course it is a bit of a leap to go from changing memory for a traffic sign to getting someone to falsely believe that they witnessed an entire event. I became eager to figure out how to do this so I could see it happen before my eyes. And I spent considerable time talking with students and colleagues to devise a way of planting an entire event in someone's mind for something that would have been at least mildly traumatic if it had happened. Eventually the idea came. I would try to get people to believe that they had been lost in a shopping mall for an extended time, that they were upset and crying, and that they were rescued by an elderly person and reunited with the family. In the formal experiment, we gave subjects life stories ostensibly obtained from their family members; some stories were true, but the one about being lost in the mall was false.[86]

The results surprised even Loftus:

We showed that you could get people to believe that they were lost in a shopping mall for an extended time, frightened, crying, and eventually rescued by an elderly person—this completely made-up scenario—we just planted it in the mind of our subject after three suggestive interviews, and then I got to watch other investigators pick up this 'lost in a mall' paradigm and plant even more bizarre and unusual memories.

It's worth emphasizing that extreme techniques of persuasion were not used. This was not brainwashing. The subjects were certainly lied to, for the purposes of the experiment. But it was the sort of lying, perhaps based on sincere belief, that can happen all the time in real life.

The distinguished Swiss psychologist Jean Piaget told a story which has become part of the folklore of childhood memory research:

'I was sitting in my pram,' he recalled, 'which my nurse was pushing in the Champs Elysées, when a man tried to kidnap me. I was held in by the strap fastened around me while my nurse bravely tried to stand between me and the thief. She received various scratches, and I can still see vaguely those on her face. Then a crowd gathered, a policeman with a short cloak and a white baton came up, and the man took to his heels. I can still see the whole scene, and can even place it near the tube station'.[87]

In fact, the kidnapping never happened—his nursemaid confessed thirteen years later that she had made the story up. But she was an authority figure and so were Piaget's parents and they often discussed the story with him as he grew up, which was enough to create a vivid and detailed memory of the event.

If a psychotherapist believes, as Lenore Terr seemed to, that psychological symptoms, dream images, confused emotions, are the signs of unremembered child abuse, she may convey this to the client as a truth, just as Loftus and her team passed on the story to the subjects that they had truly been 'Lost in a Mall'.

Loftus's famous experiment has been repeated many times and with different variations. But the first thing to say is that this is not a miracle procedure. It does not produce 'Manchurian Candidates' whose entire memories have been erased and new ones substituted. In Loftus's experiments, about 25 per cent of the subjects came to believe that they had been lost in a mall and rescued. Which means that 75 per cent, confronted with the psychological pressure of having to deny something that a near and dear relative apparently said was true, nevertheless held out for the reliability of their own memories.

Robyn Fivush, who has studied how children's autobiographical memories are shaped by parent–child interaction, was more cautious than Loftus about the creation of false memories in the laboratory:

Beth Loftus has a very different view from mine. She thinks you can get anybody to remember just about anything.... I would phrase it as you can get some people to remember some things under the right conditions. Memory is reliable for the gist. We do forget or confuse details. There's that old song in Gigi—'Ah, yes I remember it well—you were dressed in green. No, I was in blue'. But they *do* remember they had this wonderful romantic dinner together, and they understand what that dinner meant, they share that together. But she might have been wearing a blue dress or a green dress.

Loftus's false memory results have also been criticized for being based on just a small number of subjects and applied to a much wider range of situations than are justified. Looking back over fifteen years of such research, Loftus is untroubled by the criticisms:

There were 24 subjects. Believe me, articles have gotten in *Science* and *Nature* with fewer subjects. But very quickly, other investigators seized upon the methodology—'we've talked to your parents and we've found out some things that have happened to you—try to remember them if you can'— and addressed a lot of the early criticisms. One more common criticism is 'getting lost is so common, show us that you can do it with something more bizarre and unusual'. So some investigators produced a false memory that when you were a kid you had an accident at a family wedding, or somebody else used the story that when you were a kid you were a victim of a vicious animal attack. With another researcher it was that when you were a kid you nearly drowned and had to be rescued by a lifeguard. And so we have now many studies out there by other scientists who've got much higher rates of false belief and false memory than we did because they're getting better at doing it.

The significance of these experiments is not that nothing anybody says can be believed. But it does establish that we cannot any longer say, when presented with a detailed and convincing 'memory' of something that happened in someone's childhood, that it must be true. It might be, of course, but it needn't be, based solely on someone's say-so. And, indeed, our own memories of dramatic events from our childhood need to be questioned, too.

The Franklin murder case centred on an unusual example of recovered memory in that it dealt with a memory of observing the abuse—the murder, in fact—of another child. But while that case was going through the courts another lawsuit was being prepared that was to bring the topic of repressed memories into a much wider public arena. Whatever the truths of the Franklin case, it concerned the murder of a child, an event that most people never have to experience or contemplate. But the publicity given to the Holly Ramona case in the early 1990s dealt with allegations of child sex abuse, something that—regrettably—affected a much wider group.

In this case Holly Ramona, a young woman with bulimia, went for psychotherapy and was told by the therapist that 70 to 80 per cent of bulimia patients had been sexually abused. Ramona had no memory of abuse at the time but when she also admitted a fear of snakes and a history of urinary infections, that clinched the diagnosis for the therapist, and after several months of therapy, combined with the so-called 'truth drug', sodium amytal, Ramona produced vague 'visions' that she and the therapist interpreted as being raped by her father. These 'revelations' led Holly Ramona to sue her father. In the course of that lawsuit, eventually dismissed, she had to release her medical records to her father's lawyers, which was to have entirely unforeseen consequences. In an interesting twist, Gary Ramona sued the therapist and a psychiatrist for the harm that had been done by their therapy. His wife had divorced him, his daughter was estranged from him, and he had lost a highly paid job.

In this suit, Elizabeth Loftus was also called as an expert witness. Unlike in the Franklin case, her views helped to achieve a verdict in favour of Gary Ramona and against the idea that Holly had repressed memories of a series of sexual abuses, some involving a dog, for more than ten years, as some kind of coping mechanism. Looking back at the case, Loftus wrote:

When I hear the story of Holly Ramona who claims that her father raped her numerous times between the ages of five and sixteen, forced sex with the family dog, and so on, and she goes into therapy at eighteen with no recollections of this happening, I think 'something's going on here, normal forgetting and remembering doesn't work this way.' If those things had really happened to her, there would have to be severe mental disturbance but this is a functioning college student who happens to have an eating disorder. This defied everything I knew to be true about memory, yet forced Gary Ramona into horrifying and expensive litigation. My role in this case was to address this question: If Holly's memories were not real, where could they come from? I would explain how strong suggestion can create false memories and how the strong suggestion to which she was subjected may have done so in her case.[88]

The Franklin and Ramona cases were only two of many examples of so-called repressed memories that made headlines in the 1990s. And for every case that reached the courts there were dozens, if not hundreds, of young women who came to believe that they had been abused as young children, having had no recollection of the abuse until being told that childhood memories of abuse could be repressed and then recovered.

7

....

The Limits of Belief

'Believe the child' is a mantra that is often used in situations where children's evidence conflicts with adults'. Frequently this occurs in criminal cases relating to child abuse, where there is an extra factor that predisposes a court to believe the children—that the adult who is contradicting the child's testimony could be doing so to avoid being convicted of a crime.

But there is a body of research carried out in the last twenty years that shows that 'Believe the children' is a dangerous injunction, because children can be wrong. These researchers are not saying the opposite—'*Never* believe the children', but instead 'You can't *always* believe the children'.

We've seen already, perhaps even in our own cases, that the memories we have of our childhood are sometimes inaccurate and can be entirely false, depending as they do on the way we have retold or discussed our childhood in the intervening years.

But the research results that have shown the unreliability of 'Believe the child' as a guiding principle were not disseminated in time to save some innocent people from imprisonment. A number of cases in the United States and the United Kingdom in the 1980s and 1990s led to convictions being obtained solely on the basis of statements made by children, with no corroboration and often flying in the face of logic and reason. These statements had been extracted from children by techniques which research has shown to be counterproductive—in trying to get at the truth, they effectively create falsehood.

Here is how a therapist in a criminal trial in the 1980s described questioning children about possible abuse: '[For] some of the kids, what we do here is we try to improve their memory and we try to unlock their brain. Sometimes when you're real scared, your brain gets locked up. You honestly don't remember some stuff. It gets stored right back here in the back of your filing cabinet in your brain under "Z".[89]

This therapist had been responsible for obtaining stories of abuse from young children in California, part of a notorious investigation which led to criminal charges against the people who ran the McMartin Daycare Center in Manhattan Beach. Bumper stickers saying 'We Believe the Children' became popular, as mass hysteria took hold and parents interrogated their children about possible abuse while they were at nursery school. In this particular case, the allegations started after a mother questioned her son about some redness she noticed on his bottom. The child was two years old and could not yet speak in complete sentences, but the mother—later diagnosed as a paranoid schizophrenic—convinced herself that a school aide, Ray Buckey, had sodomized the child. She also claimed to the police that Buckey's mother had been involved in satanic practices, and had taken the child to a church where he was made to watch a baby being beheaded and forced to drink its blood.

The day after the child's mother called the police, a letter was sent to parents of 200 children at the day centre, asking the parents to question their children for evidence of 'oral sex, fondling of genitals, buttock or chest area, and sodomy, possibly committed under the pretense of "taking the child's temperature"'. The letter went on: 'Also photos may have been taken of children without their clothing. Any information from your child regarding having ever observed Ray Buckey to leave a classroom alone with a child during any nap period, or if they have ever observed Ray Buckey tie up a child, is important'.

As a result, questioning by worried parents followed by two-hour interviews by a social worker, Kee Macfarlane, produced the following allegations from toddlers at the school or older children who had attended earlier:

> Several children reported being photographed while performing nude somersaults as part of the Naked Movie Star Game.

106

Others testified to playing a nude version of 'Cowboys and Indians'—sometimes with the Indians sexually assaulting the cowboys, and sometimes vice versa.

Children testified that sexual assaults took place on farms, in circus houses, in the homes of strangers, in car washes, in store rooms, and in a 'secret room' at McMartin accessible by a tunnel. (No room or tunnels were ever found.)

One boy told of watching animal sacrifices performed by McMartin teachers wearing robes and masks in a candle-lit ceremony at St. Cross Episcopal Church.

One boy said that the McMartin teachers took students to a cemetery where the kids were forced to use pickaxes and shovels to dig up coffins. Once the coffins were removed from the ground, according to the child, they would be opened and the McMartin teachers would begin hacking the bodies with knives.[90]

As a result, seven people connected with the school, including Ray Buckey, were put on trial. After two trials in which the juries acquitted several defendants and failed to agree on Ray Buckey, the case was abandoned and Buckey released, having spent five years in jail.

One of the prosecutors (not the defenders) suspected that the specific interrogation methods used on the children had produced fabrications and fantasies, saying, 'Kee MacFarlane could make a six month old baby say he was molested.' The trials took seven years and cost $15 million, and apart from ruining the lives of the accused had left a trail of badgered and bewildered children.

It's worth pausing here to clarify exactly what such a questioning session is like. From accounts by accusers and prosecutors, it's easy to get the impression that there is a kind of civilized conversation in which a question is asked in a neutral way—'Did anything happen to you at school?'—and a child answers in a coherent and detailed way—'Well, yes, it's funny you should say that. My teacher took me into the bathroom, took down his trousers, and did unspeakable things to me. Let me describe it step by step...'

Court records of another case, similar to the McMartin case, revealed the questioning techniques that can give rise to bizarre and often implausible allegations. In 1985, in what was known as the 'Wee Care' case, after the nursery school where the allegations arose, a four-year-old child having his temperature taken rectally told a nurse, that 'That's what my teacher does to me at nap time at school'. As a result of an investigation set off by this remark, the teacher, Kelly Michaels, was convicted of 115 counts of sexual abuse and sentenced to forty-seven years in prison.

Looking back at the case from 1997, the prosecutor in the Wee Care case conceded that the interviews of the children were 'horrible', but that no one knew, way back in 1985, how to interview small children. 'It was a completely gray area back in that time,' claimed the lawyer.[91]

Today, it doesn't take a trained forensic interviewer to realize that these interviews were 'horrible'. Anyone glancing at the transcripts of the interviews would realize that they fall far short of making a case of sexual abuse 'beyond reasonable doubt' against Kelly Michaels.

Take one small extract from an interview which led to the allegation that Kelly Michaels had put a fork in a child's anus:

INVESTIGATOR: Did she put the fork in your butt? Yes or no?

CHILD: I don't know, I forgot.

INVESTIGATOR: Oh, come on, if you just answer that you can go.

CHILD: I hate you.

INVESTIGATOR: No you don't.

CHILD: Yes I do.

INVESTIGATOR: You love me I can tell. Is that all she did to you, what did she do to your hiney?

INVESTIGATOR (2): What did she do to your hiney? Then you can go.

CHILD: I forgot.

INVESTIGATOR (2): Tell me what Kelly did to your hiney and then you can go. If you tell me what she did to your hiney, we'll let you go.

CHILD: No.

INVESTIGATOR: Please.

CHILD: Okay, okay, okay.

INVESTIGATOR: Tell me now, what did Kelly do to your hiney?
CHILD: I'll try to remember.
INVESTIGATOR: What did she put in your hiney?
CHILD: The fork.

This is very far from the quiet, unemotional, and objective questioning that is now required of police and social workers in such situations. First, there is ample evidence from the first question that the 'right' answer for the child to give is that Michaels put a fork in the child's anus, but it takes eight more pressured questions to get the child to give that answer. In the course of those questions, the child denies that she remembers any such thing, but this denial is rejected. (What happened to 'Believe the children?' here?)

Instead of winning the child's confidence, the interviewer creates hostility—'I hate you'. Then—and this is something to be found in countless such interviews of the 1980s and 1990s—the interviewer threatens not to let the child go until she gives the 'right' answer—'What did she do to your hiney? Then you can go'; 'Tell me what Kelly did to your hiney and then you can go. If you tell me what she did to your hiney, we'll let you go'. This is clearly a lose-lose situation for the child. If she doesn't say that Michaels did something to her 'hiney' she'll stay there for ever.

When such transcripts are presented in court or in trial transcripts, they are often missing half the picture, literally. I was peripherally involved in a case where two nursery nurses in the UK had been accused of sexually abusing a number of children, and I helped their legal team by editing clips from many hours of videotaped interviews with children. Among the many accusations obtained by high-pressure questioning was that one of the nurses had put a saw into a child's vagina. It was a toy plastic saw but nevertheless not the sort of thing that could be accommodated without pain or extreme discomfort. Here's part of the transcript of the interview. ('Debbie' is not the child's real name.) The allegation emerges at the end.

MOTHER: Can you listen to Helen [a police officer] because it's very important?

HELEN: Right, there, there's nothing to worry about. Okay?

MOTHER: So can you tell her all again what happened? Don't be frightened.

CHILD: Why have you stopped coming in and out?

HELEN: Why? Because I know you need to tell me something so I'm sitting quite still so I can hear you say it and then I'll stop asking it.

MOTHER: And we won't ask any more; you just say it once and then that's it because you've told Helen—

HELEN: Shall I go, shall I, because I don't think you want to talk to us?

MOTHER: Debbie—

CHILD: Why have you—

MOTHER: Debbie—

CHILD:—got two watches?

HELEN: That's my watch and that's the clock on the wall.

MOTHER:—Debbie, you've told Helen now, haven't you?

CHILD: And has, ehm, Vanessa [another social worker] got one?

HELEN: She has.

MOTHER: Debbie . . .

HELEN: She, she was wondering—when I was back through there before, she says, 'Ask Debbie, will you, if anybody's touched her nunny?' So I've asked you. [Noise from outside room]

MOTHER: And you've told Helen now and—

CHILD: Who's that banging?

MOTHER:—nothing's happened to you, has it?

CHILD: Who's that banging?

HELEN: Can you tell me—

CHILD: What's that banging?

MOTHER: It's—

HELEN: It's somebody downstairs at work. How did—what did he touch your nunny with?

CHILD: . . . saw-saw.

HELEN: With a what?

CHILD: A saw.

MOTHER: Don't be silly.
CHILD: Yes.
HELEN: A saw?
CHILD: nods.

In the transcript, the saw comes out of nowhere, just as it might if the child suddenly remembered a real event. But it is clear from looking at the video of this interview that the child has been playing with a plastic saw, using it to cut play dough, all the time she is being questioned. Isn't it likely that, in sheer desperation, in order to bring the tedium to an end and give these adults something that they obviously want, the child chooses the nearest object to hand and makes that what the nurse put in her 'nunny'? There are other examples in these interviews where watching the videos provides a very different interpretation of the words in the transcript.

It is significant that the 'expert' who appeared on behalf of the parents who claimed abuse, admitted in court that he had not viewed the videos of the interviews but based his view that abuse had taken place merely on reading the transcripts. In this case, the charges were eventually thrown out, but only after one of the accused nurses had spent three months and the other nine months on remand in prison.

In these cases, some of the allegations are either so extreme or, in some instances, impossible that some other explanation is necessary. Whatever had happened at the Wee Care Day Center, we can be pretty sure that Kelly Michaels never 'put a car on top of' one child without being observed or without any sign of serious injury, or that another child had been turned by her into a mouse. But such improbabilities were not enough to outweigh the 'Believe the children' attitude that resulted in a guilty verdict and the forty-seven-year prison sentence. The verdict in the Wee Care case was overturned on appeal, after Michaels had been in prison for five years, eighteen months of that in solitary confinement for her own protection.

Just as with the Franklin and Ramona cases, the eventual overturning of the verdict was based on evidence from experts that the testimony for the prosecution could well have been wrong, the result of mistaken beliefs by witnesses, and of suggestive questioning by police or other investigators. After the Franklin case, Elizabeth Loftus devised experiments to show how mistaken or false memories could be created. One of the expert witnesses in the Wee Care case was Maggie Bruck, a psychologist at Johns Hopkins University in Baltimore, and she went on to carry out a series of experiments with a colleague, Steve Ceci, which demonstrated what had until then been a hypothesis—that merely by the way you question a child you can create false memories which are dramatic and enduring. They began by defining the interviewing techniques they believed could produce false statements from young children:

Interviewer bias characterizes those interviewers who hold *a priori* beliefs about the occurrence of certain events and who mold the interview to maximize disclosures that are consistent with those prior beliefs. One hallmark of interviewer bias is the single-minded attempt by an interviewer to gather only confirmatory evidence and to avoid all avenues that may produce disconfirmatory evidence. Thus, biased interviewers do not ask questions that might provide alternate explanations for the allegations or that might elicit information inconsistent with the interviewer's hypothesis. In addition, biased interviewers do not challenge the authenticity of a child's report when it is consistent with their hypothesis. Even when children provide inconsistent or bizarre evidence, it is either ignored or interpreted within the framework of the biased interviewer's initial belief. In contrast, when the child's statement is incongruent with what the biased interviewer believes, it will be challenged or pursued with repeated questions designed to align the child's subsequent reports with the interviewer's initial beliefs.[92]

Clearly, in the brief extracts above, there is evidence of interviewer bias. It is often the case in police interviews that the questioner 'knows' what happened and is seeking evidence to prove it.

But in the 1980s it had not been obvious—or at least wasn't proven—that a child would actually change his or her evidence as a result of such questioning. Then, researchers in the 1990s began to produce results which built up a case for the dramatic effects of interview techniques on the answers that were given by children.

In one experiment, children were visited at their day-care centre by a stranger. A week later, they were questioned in two groups. All were asked leading questions, but in one of the groups other suggestive techniques were used as well, such as peer pressure ('The other kids said that . . . '), positive consequences (praising certain answers), negative consequences (telling the child that this was not the right answer and repeating the question), suggestions to think about questions to which they had replied 'no', and encouragement to speculate ('tell me what might have happened').

With merely leading questions, such as 'What was his moustache like?' when he had no moustache, 83 per cent of the children gave accurate answers to questions about the visit by the stranger. When combined with the other techniques, the number of correct answers went down to 42 per cent. And these were not exotic techniques—at the time of the research, transcripts of interviews with children in connection with child abuse cases were full of examples of such methods.

Of course, the people who use these techniques are not out-and-out villains, at least in most cases they are not. Even if they believe that, for example, a particular child has been abused they are not trying to prise a false statement out of the child. They argue, however, that children are often afraid or ashamed to speak of the abuse and that special techniques like those mentioned, as well as visualization of what *might* have happened along with role play with dolls, are important to unlock the truth.

Maggie Bruck and Steve Ceci wanted to develop this research to compare the *risks* of restricting interviewing techniques—and not getting at sensitive and sometimes sexual material—with the *benefits* of minimizing the number of false answers. They also wanted to know whether there were any characteristics of false memories created by suggestive questioning that could be used to distinguish them from genuine memories.

This Bruck–Ceci experiment involved 16 children aged from $3\frac{3}{4}$ to 5 years old at two day-care centres in New York. To discover whether children's memories responded differently to positive events compared with negative ones, the experimenters questioned the children about four 'events', two of which had really happened at the children's day-care centre, and two of which had not. Of the true events, one was pleasant and the other unpleasant and the same was true of the false events.

Here's how it worked:

The true positive event was staged for each child, who met a visitor and was asked to help carry her books. The visitor appeared to trip and hurt her ankle, and the child was asked to go for help.

The true negative event was different for each child and involved some incident that had actually happened that led to the child being punished.

The false positive event was the suggestion that the child had helped a lady find her lost monkey.

And the false negative event was the suggestion that the children had witnessed a man stealing food from the day-care centre.

The children were subjected to five interviews following the events. In the first interview, they were asked if the event had happened and, if so, to provide a full account of it. No suggestive questioning techniques were used.

But interviews 2 and 3, one and two weeks after the first interview, used much more suggestion. Here's an example of how the interviewer discussed the false monkey event with the children, as described in the research report:

Those kids told me some neat stuff that happened to them. They told me that one day they were in the park and some lady came up to them and told them she had lost her monkey in the park. She asked Mary, Martha and Steve to help her find her monkey. Have you ever had anything like that happen to you?[93]

Children who accepted that the monkey event had happened were asked for a full account ('Tell me everything that happened'), and then asked six specific questions, to help the child think about the event and to provide information that might be repeated in subsequent interviews.

In interview 4, a puppet called Sedrick was used to question the children in a suggestive way:

First, the child was told that Sedrick had been talking to the other children about the event (e.g. helping a lady find a monkey in the park) and Sedrick now wanted to hear about the time it happened to this child. If the child denied it had happened to her, then Sedrick acted 'sad' and wondered why the child did not want to talk. Sedrick asked the six specific questions (by whispering these to the interviewer, who, in turn, repeated them to the child). Again, if children denied participation, they were asked to pretend and to answer the questions. Three of the six questions were repeated from interview 2, and the rest were repeated from interview 3. At the end of the questioning about each event, Sedrick praised the child for their information.[94]

Finally, in the fifth interview an unfamiliar interviewer asked each child to describe the events. Here, of course, was the meat of the experiment, a chance to find out with neutral and objective questioning what each child actually believed had happened with each of the four events.

The report of the results says: 'The suggestive interviewing techniques produced high rates of assents for both the true and false events. Although there were initially some differences between assent rates for true and false events, after only two suggestive interviews (i.e. the third interview), assent rates were similar for true and false events as well as for pleasant and unpleasant events'.

Since part of the purpose of the experiment was to weigh the merits of arguments for and against suggestive questioning, the researchers' detailed analysis of children's responses during the five interviews came up with a very interesting result. The experimenters describe suggestive interviews as a 'double-edged sword', on the basis of their discovery that, as supporters of memory recovery have said, these techniques *did* persuade children to talk about the true negative event when they were initially reluctant to do so. However, as suspected, they also promoted high rates of false memories for events that had not occurred.

There is one final insight from this work that is relevant to the use of children's testimony in court settings. Using videotapes of deliberately suggestive interviews with children, Bruck and Ceci showed that professionals, such as social workers and police investigators, did no better than chance at discriminating false from true reports. But, in

115

fact, the experimental analyses did show some distinguishing characteristics between true and false accounts. For example, the children's
accounts of false events contained more details and more spontaneous
utterances than the narratives for true events. They also contained
more temporal markers, and more elaborations, such as adjectives,
adverbs, and metaphors. Both of these results are quite surprising and
certainly go against the 'common-sense' view, expressed by some
researchers, that detail, elaboration, and spontaneity are indications
that someone is remembering a real event.

It was in an entirely different research project, involving people who
believe the impossible every day, that Richard McNally's collaborator,
Susan Clancy, she of the 'hot tub and hot ski instructor' false memory,
found confirmation of how persuasive a false memory can be to the
people who hold it and the people who hear it. It all began with UFOs.

In 1962, some people believe, aliens from a far-off world began to
invade the airspace of the Earth and abduct citizens, taking them away
in their spaceships and conducting medical and physiological experiments on them. Sometimes, these experiments involved having sex
with the Earthlings, or subjecting their reproductive organs to detailed
analysis. More and more Earthlings, usually Americans, came forward
to describe their experiences in dramatic and emotional terms, often
re-experiencing the terror they felt inside the alien spaceships, and
painting vivid word-pictures of the beings who were so cruelly mis-
treating them.

A detailed account of the UFO abduction phenomenon was given in
a book by a Harvard psychiatrist called John Mack, described as a
'Pulitzer prize-winning Harvard psychiatrist'. (He won his prize for a
biography of Lawrence of Arabia, in 1977.)

When the book was published in 1994, the publisher, Scribner, not
previously known for promoting nonsense, claimed that the alien
encounters were all 'real experiences' and 'above all, authoritative'.

For the purpose of describing the work carried out by Richard
McNally and Susan Clancy on UFO abductees, I am going to assume—as they did—that these experiences did not really happen, at
least as described by the 'abductees', and follow a line of argument

that shows common characteristics shared between the 'memories' of abductees and the recovered 'memories' of some people who believe they experienced child sexual abuse. I realize that there is a small chance that I am dismissing what, if true, would be one of the defining experiences of human civilization but I am prepared to take that risk.

The research carried out by McNally and Clancy at Harvard began when John Mack, then a psychiatrist at the same university, was investigated by a panel at the Harvard Medical School for possibly implanting false memories of UFO abductions by hypnotizing people who thought they might have had that experience. Mack was eventually cleared by the panel after other academics, including Alan Dershowitz at the Law School, felt that it would be a violation of the principles of academic freedom to stop a Harvard academic propagating the idea that UFOs came down to earth and abducted Americans. But Mack was encouraged to enrol other scientists in his research—not as abductees but to provide extra input into analysing what exactly was going on with the subjects.

This was where McNally came in.

'What was so interesting about this,' McNally told me, 'was that at that time we had a problem with the subjects with recovered memory of sexual abuse—we could corroborate none of their memories. But of course merely lacking corroboration doesn't mean that it's a false memory, and so it was very, very difficult to tell how many of these were what I would call a "genuine" recovered memory that was neither traumatic nor repressed. So we were thinking what we need to do was get a group of folks who we were pretty convinced had *false* recovered memories of trauma.'

McNally told Mack that he was going to search for these 'false' memories by putting an advertisement in the local paper saying 'Have you been abducted by aliens?'

Mack said to me 'Rich, if you put an ad in the paper you're not going to get real experiencers, real abductees.' I said 'what do you mean not "*real* ones"?' He says 'You're going to get people calling up, playing jokes or whatever.' Boy, was he right! We had some real characters. We had this one guy, every day for about two weeks, on the lab phone overnight, and we'd play it and we'd get "Eeeehooorghuuueeaaaoiiiough", sounds like R2D2, like some kind of space alien. And this went on every day for a week.

The researchers also had a recorded message from a salesman at a local Chevrolet dealership who said he had been abducted by aliens. Susan Clancy rang and spoke to the man, saying that she was replying to his message about their study. McNally told me what happened next:

He says 'Study? Study?' and she goes 'Yes, you had said that you had been abducted by space aliens and you were interested in this project', and he goes 'Space aliens? What are you talking about!' and at that point you can hear all the other salesmen at Boston Chevrolet laughing at the other end as the real Bob Smith slams down the phone. But we actually did get our 'real' abductees, as John Mack would put it.

When memories of childhood abuse are 'recovered' from psychologically ill patients, many therapists believe in the reality of the memory because the recovery is accompanied by such emotional intensity.

'Our study indicates the fallacy in this argument,' McNally said. 'The UFO abductees we studied evinced objective signs of emotional, psycho-physiologic reactivity while recollecting memories of traumatic events that almost certainly had not happened'.[95] (I think that 'almost' is unnecessarily cautious.)

'In fact,' McNally went on, 'they're at least as reactive as the PTSD patients. My interpretation of that, of course, is that it shows the power of imagination, so that if you actually believe you've been traumatized you're going to show physiology consistent with that belief. Now, of course that's not how the international space alien abductee community interpret it. John loved our study. He saw the results before it was published, and he said there's no way this can have internal origin, meaning imagination, the results are so powerful. I said "John, you underestimate the effects of imagination"'.

It's perhaps not surprising that reliving an imaginary experience can have the psychological and physiological accompaniments of real traumatic memories. Apart from anything else, most of us have had vivid dreams that lead to strong emotional reactions.

The McNally–Clancy study showed that another common factor with the UFO abductees and people with recovered memories was the way in which the earliest memories of the abduction experiences had emerged. First, the two types of experience—UFO abductions and recovered memories of childhood abuse—seemed to be triggered by specific well-publicized events. The recovered memory movement started in the late 1980s and 1990s, following the publication of *The Courage to Heal* and a couple of similar books, along with media coverage of trials like the Franklin and Ramona cases. In the same way, people's memories of being abducted by aliens only started after a spate of books and television dramas on the topic in the early 1960s, and they increased after the publication of John Mack's book, whose credibility was enhanced by his Harvard professorship. And the abduction memories often described similar types of aliens to those dreamed up by screenplay writers and movie art directors.

One 'abductee' described to Susan Clancy how he had first come to believe that he had been abducted:

It was around the time I hit puberty that everything changed. For some reason, I started to have images of aliens popping into my head. Did you see that movie *Signs*—the one with Mel Gibson? The aliens looked more like those, not like the more typical ones. I'd be walking to school and then POP—an alien would be in my head. Sometimes I'd hit my fist against a wall, because then the pain would help me think of something else. I really thought I was going crazy. After a while, I told a friend . . . He gave me a book to read. The book was called *Abductions* and it was written by a famous psychiatrist at Harvard. It had lots of stories about people who'd been abducted. I read the book in one sitting. I couldn't put it down—it all clicked with me. I knew what the people were going to say even before they said it. I completely got what they felt—the feelings of terror and helplessness. I couldn't stop thinking about the book. Once I started thinking maybe I'd been abducted, I couldn't stop. Finally I told my therapist about what was going on, and she said she couldn't help me with this, but she referred me to a psychologist in Somerville, someone who worked with people who believe in things like this. The first time I went to see him, he asked me why I was there. I opened my mouth to talk, but I started crying and couldn't stop.[96]

This account raises the second point of similarity between UFO abductees and recovered memory patients. Most UFO abduction memories emerged after therapy sessions with therapists who believed in the existence of aliens and who had a checklist of 'signs' to look for, that guaranteed that the client had been abducted. Here's one therapist quoted in Mark Pendergrast's book:

There's a ten-point checklist which cues me to look out for these cases. That includes things like missing time at night, nightmares about UFOs or vampires. Very often, people who have been abducted will wake up at the same time every night. The subconscious remembers everything; it's trying to protect itself. Some people are so vigilant and don't know why. They wake up and can't sleep until dawn. Sometimes they experience bodily sensations, tingling, paralysis, or pain. Some have mysterious marks on their bodies, bruises, scars, when they wake up. I've seen these. A lot of UFO abductees tell me they had ear problems as a child, which is a sign of an implant which can't be detected by our science. The aliens put them in to observe us.

Another sign is if someone reacts violently to the subject of UFO abductions. The other day, I was giving a public speech. As soon as I started talking about UFO cases, one woman just shot out of her chair and ran out of the room. I almost stopped the speech to run after her, because I'm sure that she was taken up by a UFO and needs to face it.[97]

Ludicrous as they might sound, these are the beliefs of someone who has been allowed to call herself a psychotherapist, set up an office, and offer psychotherapy to trusting and often mentally ill clients, who had no reason to doubt what this authority figure was saying. Here's one 'abductee's' description of the therapy process:

It was . . . common for us to seek [memories] out where they were—buried in a form of amnesia. Often we did this through hypnosis. And what mixed feelings we had as we faced those memories! Almost without exception we felt terrified as we relived these traumatic events, a sense of being overwhelmed by their impact. But there was also disbelief. 'This can't be real. I must be dreaming. This isn't happening.' Thus began the vacillation and self-doubt, the alternating periods of skepticism and belief as we tried to incorporate our memories into our sense of who we are and what we know.[98]

There's a third point of similarity between the UFO abduction stories and recovered memories. People would rather believe in an explanation for their symptoms, however implausible, than live with doubt and puzzlement about why they were suffering the way they were.

'Everybody I spoke with had one thing in common,' Susan Clancy wrote. 'They'd begun to wonder if they'd been abducted only after they experienced things they felt were anomalous—weird, abnormal, unusual things. The experiences varied from person to person. They ranged from specific events ("I wondered why my pajamas were on the floor when I woke up") to symptoms ("I've been having so many nosebleeds—I never have nosebleeds") to marks on the body ("I wondered where I got the coin-shaped bruises on my back") to more or less fixed personality traits ("I feel different from other people, a loner—like I'm always on the outside looking in"). Sometimes they included all of the above. Though widely varied, the experiences resulted in the same general question: "What could be the cause?" In short, it appears that coming to believe you've been abducted by aliens is part of an attribution process. Alien-abduction beliefs reflect attempts to explain odd, unusual, and perplexing experiences'.[99]

Richard McNally tells a story that illustrates the strength with which people cling to *any* explanation rather than being left uncertain. At a meeting organized by John Mack, people who believed they had been abducted were encouraged to bring in any physical evidence of the abduction, to be analysed by physicists or chemists. One of the physicists, Professor David Pritchard, said to Richard McNally: 'There's this guy who comes in and says "what I have here in this phial is this tiny wire. This is actually an alien probe that I extracted from my rectum"'.

McNally took up Pritchard's account:

Now, Pritchard didn't tell me what this guy was doing extracting things from his rectum, we'll just let that pass, so he says 'We'll give it to our chemists to analyse it, see what the constituents are.' In any event this guy comes back, and they've finished analysing it and Pritchard tells him 'This is basically good news, this is not an alien implant at all. As a matter of fact, we've determined what this wire is. This is not a wire at all, it's actually a thread from your underpants that has been hardened and encrusted with the remnants of an old haemorrhoid . . .' Pritchard can do this with a straight face. So how did this guy respond? Does he say 'Oh, boy, thank you, Professor Pritchard for relieving me'? No. What does he say? *'Chemists can make mistakes too!'* That's how these guys think.

It may not seem strictly relevant to the science of memory to dwell so long on a topic which most people will see as a fringe phenomenon. Unlike memories of child sexual abuse, the only harm UFO abduction memories do is to the psychological well-being of the 'abductees'. No aliens are falsely accused, convicted of kidnapping, and sent to prison. While the belief in a false memory of sexual abuse can alienate parent from child—a true and distressing event—the alien nature of abductors is entirely fictional and can be seen as no more harmful than an appetite for fantasy novels or video games. But these stories are relevant—they tell us about the human capacity to believe firmly in events in their personal histories which have never happened. Further, if there are people who consider themselves professionals who can be persuaded to believe in this stuff, how much easier it must be to persuade other therapists, social workers, and the police that there are such phenomena as repressed memories that can be recovered by specific psychological techniques. In the next chapter we will see how some of the beliefs of memory recovery therapists are almost as strange as those of UFO abductee therapists.

There's a sad footnote to the story of the rise and fall of the UFO abduction experience. (The aliens don't seem to be visiting nearly as much these days.) In 2004 Professor John Mack, on a visit to the UK, was run over and killed by a drunken driver. According to a close friend, Mack communicated (ungrammatically) from beyond the grave a few days later, saying of his own death: 'It was like a puff of wind lifted me up. I never knew dying could be so easy'.[100]

8

. . . .

Crimes of Therapy

In the 1990s, in the north of England, a young woman made a series of allegations of sexual abuse against two men. One was a teacher and the other was her father, so I'll call it the Father/Teacher case. In a statement to police she described incidents dating back to when she was ten in the case of the teacher, and as young as three in the case of her father. These offences had not been mentioned by the young woman to anyone before her police statement. Some of them were said to have occurred during the period when she could have been expected to have childhood amnesia, but others when she was a little older.

She gave the police a detailed account from memory of an incident in which she alleged that her father made her masturbate him. She included a lot of detail about the circumstances leading up to the incident and about her own emotional reaction to what had happened. She also described a second incident when she claimed that her father had touched her vagina, then inserted his finger into it and then progressed to kissing it. This also included much detail about the surrounding circumstances, including a claimed memory of what she and her father were wearing. Under questioning she made further accusations against her father, of later episodes of abuse, as well as similarly detailed accounts of rape and sexual abuse by the teacher, who taught at her school when she was ten. She said that he had abused her regularly and unobserved in an unlocked classroom while giving her lessons on her own.

The woman's accounts of the abuses were accompanied by vivid detail, including descriptions of the rooms where the events were alleged to have taken place, the clothes worn by the perpetrators—including other men she claimed the father had allowed to participate—and the specific acts, which included masturbation, oral sex, and rape.

The defence presented arguments about the unreliability of very early memories, and the improbability that such memories could have been repressed for such a long time, but the detailed descriptions of the events, their distressing nature, and the confidence with which the young woman told her story led to the trial and conviction of both men, who were sentenced to long terms of imprisonment.

So far, so normal—unfortunately. We live in a world where sexual abuse of children is not rare, and where, if a crime is likely to come to light, it is most likely to be as a result of the victim coming forward, sometimes years later. Furthermore, there is rarely corroboration in such cases, and the legal system deems such crimes to be so serious that convictions can be obtained on the basis of unsupported claims by the alleged victim of crimes which took place many years ago.

There are many such cases where convictions have been obtained that are justified. Men—usually—who thought they had got away with abuse of their own children through cover-ups maintained by threats or violence have finally been brought to justice years later.

What was at the heart of the Father/Teacher case, and of many allegations of uncorroborated sexual crimes, was the science of childhood memory. This is a comparatively new branch of psychology, stimulated—as I've described—by a succession of cases which appeared to involve memory processes that sometimes contradicted what was known about how memory works. Much of what we know nowadays about childhood memory has come from carefully designed, statistically valid experiments which need expert psychological research to design and interpret, so that the results are valid, particularly when they contradict 'common sense'.

When other aspects of science and technology are involved, legal systems around the world generally allow the defence or the prosecution to call experts to give evidence if their field of expertise deals with matters which the average jury member would not be expected to know about as a matter of course.

If the last case had been related to, say, a car crash caused by a severed brake fluid pipe, in the normal course of events technical experts would have been called to present scientific data about the technology of car braking systems, the likelihood of corrosion in brake pipes, the factors that might lead to the conclusion that the pipe had been deliberately cut, and so on—not matters a jury would be expected to know about as a matter of course.

Now, in cases which revolve around whether an adult would have remembered something that happened when she was a small child, or indeed whether memories of traumatic childhood events can be repressed, it might be argued that in the light of research by psychologists over the last twenty years, there are all sorts of aspects of the topic that the average jury member could not assess on the basis of his own experience. Most of us have erroneous views about the nature of our own memories and about memory in general. Is this, then, not a subject where expert witnesses could be called in to help the jury understand the complexities of the issues? According to British law, the answer is usually no.

In the Father/Teacher case, lawyers for the accused complained that they had not been allowed to present crucial evidence by a memory expert. In their verdict, the judges quoted the precedent that was usually used to rule that expert evidence was only admissible when it would help the jury understand a topic that was outside its ordinary experience. But then, in a ruling that made an exception of this particular case the judges decided that, in spite of the precedent they quoted, if an expert on childhood memory had been allowed to testify at the trial, the two men might not have been convicted.

It can sometimes seem as if the kind of research mentioned in the earlier chapters is of merely academic interest. Learning word lists, asking students about their childhood memories, interviewing UFO abductees—all of these provide interesting information about human memory, but do they really matter? Isn't much of it just abstract theorizing with little practical application?

Well, it mattered to the father and the teacher, but unfortunately it mattered too late. These two men would not have served long terms of imprisonment if expert testimony about the science of memory had been accepted and believed at the initial trial.

Their story is just one example of the way in which non-science, and sometimes nonsense, about childhood memory has crept into the courtroom and led to the incarceration of innocents.

There are clearly genuine abusers in prison today, solely based on the word of their victims. Is there a price to pay for this approach, which is unique to the crime of child sexual abuse? (Other crimes generally require corroboration.)

Studies of how jurors behave suggest that they base their assessment of witness testimony on many different factors, some of them irrelevant to the truth or falsehood of the evidence. Reporting on psychological experiments with jurors, psychologist Daniel Schacter wrote:

When confronted with a highly confident eyewitness, juries tend to focus more on that person's believability than on the original witnessing conditions that may have made it difficult for the witness to perceive or identify the perpetrator. Even though juries believe confident witnesses more than uncertain ones, eyewitness confidence bears at best a tenuous link to eyewitness accuracy: witnesses who are highly confident are frequently no more accurate than witnesses who express less confidence.

Elizabeth Loftus and a colleague looked at how a group of mock jurors were swayed by the amount of detail in a witness's statement. The inclusion of trivial detail in eyewitness accounts, such as seeing a store customer buy and then drop 'a box of Milk Duds and a can of Diet Pepsi' rather than 'a few purchases', influenced mock jurors' decisions. Comparing prosecution witnesses' statements with and without trivial details, the researchers found that jurors were more likely to select the guilty verdict if presented with the detailed account. The psychologists said that 'communicators should choose their words very carefully, because the minor details that a communicator reports might be as influential as information that has genuine significant value'.[101] (Remember the 'seductive detail' effect in Chapter 2.)

In the Father/Teacher case, the victim's accounts were full of vivid detail. What's more, if the jury believed the allegations of the earliest abuse, said to have taken place when the girl was three, they are likely to have seen those as 'confirming' the later ones, even though there was no independent evidence to support any of them. And yet, at the appeal, the evidence of a memory expert was accepted, over the objections of the prosecution, and he said that he had never come across a person who had been able to provide a detailed narrative account of an event that had taken place at the age of 4 or 5. A child of 4 might well remember something that had happened when he was 3, but by the time he was 7 or 8 he would have forgotten it and it would not be recaptured.

While it is a hopeful sign that the court believed the expert's evidence against the idea of repressed memory, memory science is not yet fully allowed into the courtroom. His evidence was very much the exception to the rule, because the judges felt that the specifics of childhood amnesia would be outside the jury's normal experience. But then surely this is also true of the mechanisms by which false memories can be created in the laboratory, or the absence of reliable evidence of widespread repression of childhood memories, or the extent to which biased questioning of children produces inaccurate answers, which are equally outside the experience of the average member of the public?

Scientific research is continually turning up unexpected findings about 'ordinary experience'. If new findings about memory, perception, judgement, trauma, and other aspects of being human are never allowed into the courtroom because these are aspects of life that are part of 'ordinary experience', we could end up in the situation that obtained a few years ago where a woman who fell off a bus was granted damages to compensate for the later birth of a child with Down syndrome, due to a chromosomal defect that couldn't possibly have been caused by injury or shock. In this case there were conflicting expert witnesses, but it could well be that the 'ordinary experience' of a jury—'surely the fall must have caused the Down syndrome'—was enough to lead to the strange verdict.

The manner in which the two men in the Father/Teacher case eventually got a sort of justice is an illustration of the poor state of knowledge among lawyers and the general public about the reliability of childhood memory. One of the extraordinary things about the convictions in this case was the fact that there were no other hints of abuse accusations levelled against either of the men, in spite of considerable efforts made by the police to trawl through the teacher's other pupils. The only link was the girl, who made surprisingly similar accusations against two entirely unconnected men.

Another recent UK case of allegations of childhood abuse also ended with the accused, Jim Fairlie, a Scottish politician, being cleared. In this case, his daughter, at the time a young mentally-ill woman, underwent psychiatric treatment after experiencing a range of puzzling psycho-somatic symptoms. When she was a child she had been abused by her grandfather. Her family knew this and wondered whether this abuse might have left its mark in some way and caused the daughter's symptoms. While in psychiatric hospital, the young woman's mental health deteriorated. She lost a lot of weight, tried to commit suicide, and experienced nightmares and hallucinations.

Dr Alex Yellowlees, the consultant psychiatrist in overall charge of her care, and a team of mental health professionals at this point started a broad range of treatments. During this process the daughter came to believe that her father had raped her when she was small, and in addition had beaten to death a six-year-old girl in front of her. What's more, the young woman 'remembered' that her father was a member of a paedophile ring of eighteen men, two of them Members of Parliament.

At this news, the whole family went into shock, and was split by bitterness and anger. Eventually, after further discussion with the young woman, the police decided that there was no basis for charging the father, and the woman herself came to realize that her accusations were unfounded and had been connected with the treatment she had undergone for her other problems.

Mr Fairlie tried to sue the local National Health Trust, NHS Tayside, for compensation for the family's ordeal but the action was dismissed because the judge decided the Trust owed no duty of care to the father during its treatment of his daughter. In other words, the Trust believed that it was not responsible for the harm caused by accusations against Fairlie of a series of heinous crimes that emerged during treatment (the Trust denied that Ms Fairlie had undergone recovered memory therapy at any time). Angered by this, the daughter, who had withdrawn her allegations against her father a few months after making them, sued the council herself for the years of distress to her and her family. In October 2007 there was an out-of-court settlement in which the NHS Authority paid the daughter £20,000 but denied any liability and did not apologize.

Each of these cases, and many others, started with a troubled young woman who believed, or whose therapy led her to believe, that her troubles were caused by bad things that happened to her in her childhood, and in the Fairlie case the woman herself came to realize that her accusations were unfounded and had been connected with the treatments she had undergone for her other problems.

Many of us believe we had a *happy* childhood. Based on our memories, which, as we've seen, can be unreliable and inaccurate, we may feel on the whole that our parents did the best job in the circumstances. We may not have got the train set we coveted one Christmas, and experienced a moment of deep sadness when our sister was given one instead; we may have suffered the loss of a beloved parent or sibling prematurely; we may have wished to live in a better-off family or a healthier one. But it is possible to have experienced all those things and yet believe that childhood was a happy time.

None of this would necessarily prevent unhappiness later in life. And yet, see what a leading psychologist, Alice Miller, has written:

The truth about childhood, as many of us have had to endure it, is inconceivable, scandalous, painful. Not uncommonly, it is monstrous. To be confronted with this truth all at once and to try to integrate it into our consciousness, however ardently we may wish it, is clearly impossible.[102]

If you accept Miller's view, there are four possibilities that might apply to you personally:

You might agree straight away, if you have suffered sexual abuse or violence, and already be trying to 'integrate it into your consciousness', using Miller's phrase.

You may be surprised by the force of Miller's statement—the *truth* about childhood, she writes, no ambiguity there—and even if you can recall no abuse you may begin to wonder about certain puzzling aspects of your own psychological health.

You may consider yourself reasonably balanced, coping well with the complexities of the modern world, and assume that you are not one of the 'many of us' Miller mentions who have had to endure this 'truth'.

Or you may dismiss what she has written as just plain wrong.

Miller has spent her life trying to understand the consequences of child abuse, and believes that all mental illness is caused by childhood trauma. It is this view in various forms that underlies the growth in recovered memories of child abuse over the last twenty years.

It is an insidiously harmful view, not because it is necessarily wrong, but because it cannot be proved to be right. Certainly, there are people who suffer from mental illness who turn out to have been abused; and there are well-balanced, mentally healthy people who were not abused. But the opposite is also true: there are people who lead happy and well-adjusted lives in spite of a history of child abuse, just as there are people—I believe—whose lives are blighted by mental illness and who were never abused as children.

But if you are depressed or anxious, suffer from panic attacks or nightmares, possibly even from schizophrenia or obsessive-compulsive disorder, life can be harsh and puzzling. 'Why me?' is the question people often ask, about physical as well as mental illness, and just as the UFO abductee with the implanted alien wire would not accept the mundane explanation offered, people who are ill are not always happy to accept as an explanation that 'stuff happens' or even that some things in life are not yet understood.

'I'm a Harvard graduate and an incest survivor,' one patient has said, 'and I have yet to hear any convincing explanation of what a survivor has to gain by this. My experience is that all of us have lost a great deal: family members we love, the belief that they were there for us, who we were as children, the image we had of ourselves. I don't know many people who would do that on purpose to themselves, just for a civil suit'.[103]

But Elizabeth Loftus said to me: 'I get asked why would anyone want to believe anything so awful as "daddy or mommy did this to them", and I can tell you, in most cases you have an excuse for all your problems, if you've misbehaved or haven't achieved as much as you should have, or you're depressed or have other symptoms, now you have an explanation. You don't have to be a bad person, you don't have to be a crazy person, you're just abused. You get bathed in a love bath by other supposed victims and victim supporters, you get sympathy, empathy, there's the benefit'.

I have tried so far in this book to present, in a very simplified form, the evidence of scientific investigations of childhood memory and how it works. But I have not only presented some of the evidence, I have also tried to show how that evidence was obtained. It is usually the result of carefully planned experiments, whose design is such that the results that are obtained can be trusted, to some level of statistical significance. It doesn't mean that they are a hundred per cent correct for all time— scientists sometimes disagree—but at least, as we've seen, such experiments are described in terms which allow another scientist to repeat the work, using the same method, and confirm the results. And if results disagree, there can be discussion of why, using the language of science.

Unfortunately, people can use the language of science without ever having used the method. Many of those who believe in the reality of repressed memories come up with statements and explanations that *sound* as if they are based on some kind of evidence. Take this statement, for example, by a psychotherapist:

Survivors never forget the trauma. It gets dissociated off. The child's ego can't hold the reality of sexual abuse and the demands of living every day. So call it something else, pretend it didn't happen, leave the body. There are lots of ways to make it not be happening. But it leaves footprints in the snow of everyday living, though the symptoms need not be dramatic.[104]

131

'The child's ego can't hold the reality of sexual abuse' has the same form as 'The bucket with a hole in it can't hold ten gallons of water'. But you can *see* a bucket and you can *see* the holes and you can *see* the ten gallons of water pouring out. Nobody has seen a child's ego holding or failing to hold anything. Similarly, 'it leaves footprints in the snow of everyday living' is a bit like 'the fox leaves footprints in the snow in my garden' but the same fundamental difference between the two statements arises.

Of course, it is possible to come up with hypothetical concepts, similar to the therapist's 'ego' and 'footprints', in real science. Electrons, black holes, antibodies, genes—all were part of scientific discourse before they could be ascribed any kind of physical reality. But there was agreement among scientists about why these concepts might exist, usually based on detailed and methodical observation, and about the methods to be used to detect them, and to exclude alternative explanations. None of that agreement exists among believers in the 'child sexual abuse' theory of psychological illness, either in terms of a causative mechanism, a methodology for proving that abuse was the cause rather than some other factor, or about the methods to be used to uncover 'memories' and treat their effects.

For Jennifer Freyd the methods of science are only one way to investigate human psychology, and she works as a scientist. Here's the opinion of a therapist with similar views to Freyd. It comes from the 'expert' testimony given in the trial of a Sunday school teacher who was accused of sexual abuse and kidnapping on the basis of so-called recovered memories, and was eventually cleared of all charges after having spent several years in prison. The therapist in the case was asked whether a diagnosis in therapy needed to be based on research-verified science. The therapist gave the following answer:

A: I think there's... there are different types of science. There's research science and...

Q: What type of science do you practice?

A: Artful science.

Q: O.K. So you're saying it's an art, it's not a science?

A: A little of both . . . I think there's a lot of counseling theory that has not necessarily been proven in scientific research.[105]

Doctors are sometimes criticized for acting on their beliefs about an illness and the best treatment for it, but the range of beliefs you will find in the medical profession about the causes and treatment of physical diseases is a model of specificity and rigour compared with the kinds of things some therapists say when treating patients they believe are suffering from the effects of repressed memories of child abuse. Here is a selection, taken from Mark Pendergrast's exhaustive survey of recovered memory therapists and their clients.

When clients tell me they have no recollection of whole pieces of their childhood, I assume the likelihood of some sort of abuse.[106]

How old do you think you were when you were first abused? Write down the very first number that pops into your head, no matter how improbable it seems to you . . . Does it seem too young to be true? I assure you it is not.[107]

How do we know if the memories are real? We use our gut reactions at times. Besides, if it takes years in therapy, that's a good indication they are real memories. I consider the vividness, clarity and specificity of the memory.[108]

I have a mental image of a map. If a person tells me they have a severe fear of genital intrusion and can't go to a gynaecologist, it fills in my map a little bit. I may tell them, 'This sometimes means something has happened, but it's not sufficient to draw conclusions. Let's put a pot on the back of the stove, and put these things in it, and see if it makes a soup.'[109]

Emotional incest can be as bad as physical sexual abuse. I think about it in terms of appropriate boundaries. A four-year-old might hear Mommy or Daddy talking about financial worries or their sex life, topics the child shouldn't be privy to. That sets the child up to be responsible for the parent. You might hear Dad talking about not getting enough sex from Mom. A parent who beats or rapes his children, everyone agrees that's terrible but emotional incest is harder.[110]

There are four classic symptoms of sexual abuse: 1. very low self-image, 2. obsessive ruminations and flashbacks, 3. disabling mistrust of men and sometimes of women, and 4. sexual dysfunction of some sort. Other symptoms? Eating disorders make me just wonder, but the research now shows that

133

up to 50% of the psychiatric population has been abused in one way or another, so it isn't surprising that a high proportion of patients with eating disorders have a history of sexual abuse. Other incest survivors may react with a startled response when people come up behind them. Some refuse to be examined by doctors, hate gynaecological exams, anything like that. There are children who don't like their throat looked at or who greatly fear going to the dentist. You just wonder if they might have been orally abused.[111]

Another thing we learned was that claustrophobia often indicates a person had had oral sex forced on them. It made a certain amount of sense.[112]

I have one current client who has retrieved memories of being sexually and ritually abused. She was working with a female therapist and already had an idea that some things had happened when she came to me. She started to retrieve memories at a rapid rate. For instance, I would pull her arm back to work on the triceps, and she would resist every time at that part of the massage. It turns out that during ritual abuse, she had witnessed another victim's arm being lifted in a similar fashion and severed at the shoulder which explained her fear. When she was four she remembered witnessing a girl her age being molested, killed, and dismembered. A lot of times, she wakes up at three in the morning and realizes, 'Oh, my God, I know what that's about now!'[113]

Reading accounts of what some of these therapists believe about the origins of psychological illness can drive one to share Mark Pendergrast's verdict on one therapist he visited: 'I came away from my interview...thinking that she was intelligent, assertive and quite possibly insane'.[114]

The theory promoted by Alice Miller and the authors of *The Courage to Heal* that all psychological illness is definitely caused by child sexual abuse has one very serious consequence—therapists who believe in this statement won't give up until they find the abuse. If you were told there was *definitely* a pot of gold worth £10 million buried at a depth of three feet in a particular field, you would go on digging until you found it. Recovered memory psychotherapy can take a long time and the expenditure of a lot of money, by the client. One patient talks of resisting four years of pressure by a therapist before she came to accept that she was an incest survivor.

One sign of the flakiness of recovered memory therapy is the fact that different therapists use different methods, and each thinks his or hers is the most effective. This may seem merely a characteristic of psychotherapy in general, but in fact genuine, science-based, psychotherapy has refined its techniques over the years and there is now general agreement among qualified therapists that the most reliably effective form of therapy for many common psychological problems is a collection of therapies called cognitive behavioural therapy, or CBT. One of the advantages of CBT is that it can be carried out over quite a short period—a few months—rather than taking up a significant amount of time, and money, over a period of years. It has also been scientifically tested, unlike many of the methods used by psychotherapists, and has been shown to be more effective for some conditions than the sort of medication that has been prescribed in the past.

Surveys of UK and US therapists with PhDs (so these are at the high end of the spectrum, since anyone can call themselves a psychotherapist and practise as one) show that 25 per cent of them believe in the existence of repressed memories of child sexual abuse and the effectiveness of a range of techniques for uncovering these memories. This belief conditions the approach they take to a course of therapy with a client who presents with any of a wide range of symptoms.

The starting point for a course of therapy is usually the patient's history. As we've seen, this is a rich and fruitful breeding ground for clues that can be interpreted as evidence of childhood sexual abuse, and leads to a situation in which the therapist is not 'discovering' abuse as a result of a history—as might occur if an adult actually describes a specific episode from childhood. Instead, armed with a belief that abuse occurred, the therapist is interpreting a wide range of normal behaviours as confirmation of that belief. If we were to assemble all the things that therapists believe are signs of early sexual abuse, from claustrophobia to eating disorders to distrust of men and so on, there would be few patients presenting themselves for psychotherapy who did not display the required signs. One patient was identified as having been abused as a child because 'she wears oversized clothing, has a makeup phobia, and experiences panic attacks during sexual intercourse'.[115]

135

A case reported by psychologist Maryanne Garry and her colleagues, from personal experience, shows how assiduously a recovered memory therapist can pursue the topic of child sexual abuse in the absence of any plausible evidence. During a seminar, a young woman who used computers a lot in her work told of going to a pain clinic because of pain and numbness in her wrists. With such a typical symptom of her occupation, it would be perverse of anyone, particularly a medical professional at a pain clinic, to ignore the possibility of repetitive stress syndrome. But, in fact, the perverse happened. The young woman was told: 'Your arms are numb because your mind is numb, because you are repressing memories of childhood sexual abuse.' Fortunately, she was aware of the repressed memory controversy and chose to reject the explanation and take her problem elsewhere.

Faced with these 'signs' that abuse occurred but with no detailed account of the circumstances or the perpetrators, many therapists work with the client, usually over a long period, to tease out the details, often piecemeal, sometimes with considerable emotion, some-times without any corroboration. But just as the diagnostic methods are unsupported by scientific evidence, so are most of the techniques the therapists use, and when some of them are put to a scientific test, it turns out that they can *create* the memories they are meant to reveal.

9

· · · · · · · · ·

'Believed-in Imaginings'

Most recovered memory therapists use a range of methods with their clients to uncover the childhood memories they suspect of causing unhappiness or illness. A survey of therapists belonging to the American Psychological Association showed that 47 per cent used dream interpretation; 15 per cent used what is called 'body memory interpretations'; 27 per cent used guided imagery; and 20 per cent used hypnosis or trance induction. And a third of them used a procedure they called 'bibliotherapy'.[116] It turns out that this consists of giving patients copies of books like *The Courage to Heal* and asking them to do the accompanying exercises and workbooks. (There is no record of patients ever being prescribed Elizabeth Loftus's books as part of a course of bibliotherapy.)

Nearly 30 per cent of therapists thought it was a good idea to refer clients who had no memory of being sexually abused to a sexual abuse survivors' group, where, I would have thought, social pressure alone might lead them to share the common perception of having been abused, particularly if the members of the group share the same range of psychological symptoms.

One patient who had no memory for abuse, and who spent $60,000 on therapy, took part in group work with others who believed they were abused. After listening to such stories, day after day, this patient began to wonder whether she too had been abused. Perhaps not surprisingly she began to have dreams in which abuse was a feature:

I would dream about my father raping me, but I would at first minimize and deny it. I dreamed about my mother sexually violating me also; she had a penis

137

and would sometimes anally penetrate me. It could have been her finger in reality, any sort of phallic thing. The idea of my mother being sexual with me is the point. . . . Eventually, I recalled that I was molested from infancy until 16. My father abused me when I was in my crib, until I was four. My father did it the whole time.[117]

Like other aspects of recovered memory therapy there is little scientific evidence to support the use of dream interpretation. The idea that some survivors experience forgotten episodes of childhood trauma only during sleep, while these episodes are inaccessible to daytime consciousness, contradicts what is known about authenticated survivors of trauma, who are troubled by regular intrusive memories of the trauma by day and by night. In addition, there is considerable evidence that the content of dreams is often derived from the events of the previous day, so a woman in an intensive programme of psychotherapy, with emotional sessions in which she is questioned about child abuse and asked to imagine episodes from her earliest childhood, would be quite likely sometimes to have dreams whose content is related to child abuse.

Nevertheless, there are numerous accounts from patients of how therapists told them that their dreams proved the reality of early abuse in their lives.

'The dreamwork became quite central to my therapy around this time,' says one patient. 'I came in with more and more dreams and nightmares. One particular dream led to my sexual abuse accusations. I dreamed I was in the same house with my parents, in a room that had hundreds of bolts up and down my bedroom door. I was quite happy with this privacy, with the door shut, with a male I felt comfortable with. Outside in the hallway, on a cold floor, my mother was sitting on a quilt. I was thinking, Why doesn't she go back to bed? There was a voice, my father's voice, shouting for my mother to come back upstairs to bed. And that was it'.[118]

Another patient had a vivid dream of being abused by her father. In her account to the therapist of the dream, she described her father cutting off the head of her teddy bear with a Swiss Army knife while he abused her.

'He's never owned a Swiss Army knife,' said the patient's mother later, 'and I found the teddy bear packed up in the basement, not a mark on it'.[119]

Such anecdotes don't prove very much, but we have to bear in mind that they are snapshots in what is usually a long-running and insistent process, in which dreams are brought to every session with the psychotherapist, and these sessions either include interpretation of the dream as evidence of abuse, or constant discussion of abuse as a cause of symptoms, and puzzlement as to why the dreams don't yet reflect this. It is an adult version of the questioning of tiny children by police or other authority figures, where the questioner will say, 'Tell me what the teacher did to you and you can go home.' In this case, it's as if the patient is being told, 'You haven't had the right dreams yet—go home and try again tonight'. And they do.

The unreliability of using dream content in this way is shown by the fact that, in one study, seventeen out of eighteen people whose children or spouses had been murdered experienced repetitive nightmares and flashbacks about the murders, even though only one of them had witnessed the crime. 'A man experienced intrusive visual images of his son's scalding even though he had not been present when the accident occurred. A woman experienced intrusive visual images of her murdered father's mutilated body, even though she had not been at the crime scene'.[120]

If there is no evidence that dream interpretation *reveals* repressed memories of abuse, there is evidence that by using these methods therapists can *create* memories of abuse that didn't happen. In experiments carried out by Elizabeth Loftus and Giuliana Mazzoni, the experimental subjects were asked questions about various events that had happened before the age of three. These included such things as being abandoned by parents, getting lost in a public place, or being lonely or lost in unfamiliar surroundings. Two weeks later, half of the subjects took part in an apparently unconnected research project about dream interpretation, in which they made notes about a recent dream and narrated the dream to a researcher.

Regardless of the content of the dream, the subjects were told that the dream was evidence that they had experienced certain events in childhood, such as being lost or abandoned before the age of three. Many of the subjects, when questioned later, came to believe that the events, that they had earlier denied, had actually occurred. The experiment was repeated with the 'interpreted' event being episodes of bullying as a child, and the subjects' confidence that the event had occurred increased after the dream interpretation.[121]

There's another sleep-related experience which is often treated by therapists as evidence of sexual abuse. It is known as sleep paralysis and is well described in the following personal accounts:

I woke up suddenly in the middle of the night. My eyes would just be open, and I would be frozen in terror in my bed, stiff. I couldn't even breathe. And I would be looking up at the silhouette of a dark figure reaching down to get me. It was really as if there were someone right there, but I couldn't move. Eventually the figure would fade completely. I would finally be able to get a breath, and I would really have to work at making myself move a finger, a hand, and finally be able to reach for a light and turn one on and sit up.[122]

Ever since I was little, I've had paralysis in my dream state, where I feel like I'm awake, but I can't move. I would see things, like another realm opened up to me. There would be incredible fear with it. It always seemed that something came into my room and touched me.[123]

I once had this terrible nightmare—at least I think it was a nightmare. Something was on top of me. It wasn't human. It was pushing into me. I couldn't move; I couldn't scream; I was being suffocated. It was the worst dream I've ever had.[124]

Sleep paralysis is normal. People who have never been sexually abused experience it. But the three accounts above were from people who, upon describing their experiences, were told that this strongly suggested that they had been sexually abused as children.

When I told my therapist about it she basically asked me if anything had happened to me as a kid. She was getting at sexual stuff, like if I had been abused. It took me off-guard, because I'd never thought about that before.

140

Anyway, she said that sometimes memories that are really traumatic get pushed down by the psyche—it's like a protective mechanism—and that they can kind of 'pop up' in dreams.[125]

Well, I had never understood this, but now I connected it with the possibility of sexual abuse, like maybe somebody came and got me in my bed at night. So I wrote in my journal about that, and I really started working on these memories.[126]

After I accepted that I was an abuse victim, it got much worse. It was constant, almost every night. Most of the time, I was frozen. I couldn't even scream. Things would be flying into my room, getting on top of me, molesting me. Some were evil-looking inhuman things.[127]

During his research into PTSD sufferers, Richard McNally came across people who had been led by their therapists to believe that their episodes of sleep paralysis were caused by sexual abuse or incest, and were relieved to discover that it was normal:

I have had people contact me who thought they'd seen ghosts, or they'd been to a therapist who said 'Oh, it's a returning fragment of a repressed memory of incest by your father', and they're doing years of therapy because the therapist didn't recognize sleep paralysis when she saw it. And this person immediately says 'Oh my God! Thank you. That's what *I* had! I was never an incest survivor.'

In her survey of 'UFO abductees', Susan Clancy came across people who believed that sleep paralysis was proof of their abduction, and were reluctant to exchange the exoticism of contact with aliens for a mundane phenomenon of the human brain. After she suggested to one such subject that his sleep paralysis was normal, he rejected this and later she heard him through the door of her office on the phone to a friend:

I swear to God, if someone brings up sleep paralysis to me one more time I'm going to puke. There was something in the room that night! I was spinning. I blacked out. Something happened—it was terrifying. It was nothing normal. Do you understand? I wasn't sleeping. I was taken. I was violated, ripped apart—literally, figuratively, metaphorically, whatever you want to call it. Does she know what that's like? Fuck her! I'm out of here![128]

Another tool used by recovered memory psychotherapists is 'guided imagery'—instructions to the client to 'imagine what it would have been like if it *had* happened'. This too has been shown to produce false memories. Two researchers, Ira Hyman and Joel Pentland, compared guided imagery for an event that didn't occur with merely thinking about such an event for one minute. They found that nearly 40 per cent of the subjects in the guided imagery group ended up with a false memory of an event which they were confident had happened, compared with only 12 per cent of the control group.[129]

One patient was encouraged to imagine an episode of abuse and write an account of it. 'I had no memory of what my father had done to me,' she wrote, 'so I tried to reconstruct it. I put all my skill, as reporter, novelist, scholar—to work making that reconstruction as accurate and vivid as possible. I used the memories I had to get to the memories I didn't have.' Her account described imagined abuse by her father when she was three: 'I lie there with his fingers crawling over me. I keep jerking, I can't help it, jerking under his fingers. I think it hurts, but I'm not sure. My flesh is soft down there, so different from the firm skin all over the rest of me. He rubs against the bars of the crib and his eyes cross and roll up behind his glasses. Suddenly he groans and slumps over the bars. His finger stops moving. Is he dead?'[130]

Another 'repressed' memory—or not—sees the light of day.

The guided imagery technique has also been used in forensic interviews of children. Children who were reluctant to give details of suspected abuse were asked, 'Imagine what it would have been like if X had done Y to you.' It's difficult enough for adults undergoing guided imagery to resist the formation of false memories, but suggestible young children find it even harder.[131]

Perhaps the best known, and most misunderstood, technique used by therapists in their search for repressed childhood memories is hypnosis. This has been used in a number of psychological and legal contexts and there is a general belief that it can extract information that is otherwise inaccessible. Richard McNally told me about an incident in Cambridge, Massachusetts, where hypnosis was used by the police to try to enhance the testimony of eyewitnesses to a crime:

There was a bank robbery in Harvard Square. A Brinks truck came up, these guys robbed it, there was a big shoot-out and a getaway car, so it was all a big upheaval. And it was a Harvard University employee who was present and saw the getaway car. And back then there were people called 'police hypnotists'.

McNally began to bounce up and down in his chair with the excitement of the story.

And so the cops brought in a psychologist, and they hypnotized this guy, and asked him about the licence plate, and under hypnosis the image comes up— it's a Massachusetts licence plate—and then he starts giving out the licence number and it surfaces effortlessly in his mind. He's convinced that's it, and he comes out of hypnosis saying 'That's it, that's it!' and the cops are very excited. So they punch in the licence plate in the computer, and sure enough there's a Massachusetts licence plate with the number!

So far, a potential success story for forensic hypnosis. But then the let-down.

'Who owns the car?' McNally said. 'President Neil Rudenstine of Harvard University! It turned out that the President of our university was not the driver of a getaway car, and when the cops asked the gentleman what he did for Harvard they found that he worked over by the parking lot and would see Rudenstine's car coming in every day, and so he was passively encoding this licence plate number and under hypnosis that plate surfaced but the source didn't. So there you have it.'

Twenty per cent of therapists who were asked about the methods they used said they employed hypnosis, although some of them appear to have a shaky idea just what it is they are using:

'I don't like all the hullabaloo that goes on around the word "hypnosis",' said one therapist, Hamish Pitceathly. 'You're hypnotized a good deal of the time anyway, such as driving on the M1 in a trance and saying, "How the hell did I get here?"'[132]

Some therapists hypnotize a patient and look for what are called 'ideomotor signals'. This rather obscure concept refers to a belief that negative life events are associated with different parts of the body. 'I try to get that "part" to respond,' one therapist said. 'And when it is ready to do so, I ask it a yes/no question. The part responds with an ideomotor signal, one finger for "yes", another for "no". We establish how old the whole person was when that part came into being. Did it come into being when the whole person was a child? Was the whole person younger than ten? Most parts, I find, are already established by the time we're five. Some are sort of put in storage as it were and triggered off by something at one stage or another'.

'Ideomotor signals' are related to what some therapists call 'body memories'. This is another concept with no basis in human anatomy, physiology, or neurology, which is presented to patients as if it were firmly founded on science. The theory seems to be that some memories are so traumatic that the brain can't cope with them and so they are 'stored' somewhere else in the body. So headaches and migraines, gastrointestinal problems, or pelvic pain, could all be 'body memories' of abuse. 'The body holds the memory. The body holds all memories,' says one therapist.

Another therapist tries to extirpate bad body memories by telling the patient to 'reclaim one square inch of her body at a time'. Patients don't always find this easy. 'If you can't find even one square inch, you're in deep trouble. Women often find it behind their knees or on their hands or in their hair. Every week, you pay attention to that square inch, like it, admire it, buy it a present'.[133]

The therapist's job is to assess from the patient's demeanour, history, and presenting symptoms where the body memory is stored, and to make an assessment of what the trauma was that caused the body memory. After being told how body memories 'work', one patient, a young woman, became breathless and felt dehydrated. 'I feel like a hot baby in a pram,' she told the therapist, and together they arrived at the 'diagnosis' that she was being suffocated by the weight of her father abusing her with his penis in her mouth.[134]

One young woman who had had polio as a child was told in a group therapy session that her illness had a 'purpose', and that she had probably been abused by her father so developed polio to 'escape' to the hospital.[135]

Another patient, told about the theory of body memories, asked for an explanation and the therapist said: 'All I can say is that the body stores the memories that the brain can't cope with. Your memories are coming out in a safe environment, a lot of inner child work is going on, so you feel safe. The body never forgets.... Trust me'.[136]

Lenore Terr, the 'expert' witness in the George Franklin murder case, believed that Franklin's daughter's habit of pulling out the hair on one side of her head represented a body memory of seeing her father kill her schoolfriend with a rock to the head.[137]

Many therapists try to regress the patient to an earlier time, in order to retrieve repressed memories. What I've described earlier in the book about research into childhood memories and childhood amnesia should suggest to most people that the idea that people can remember back to the time when they were newborns is inherently unlikely. At least, there is no evidence that this is possible. But over the last twenty or thirty years, it has been a central tenet of some psychotherapies that such early memories are preserved and can be responsible for later psychological trauma. Under therapy, the American television actor Roseanne Barr claimed that she had detailed memories of her mother molesting her in her crib when she was six months old.

One therapist described how he discovered that a particular memory came from before the patient's birth: 'I often encountered detailed descriptions of underwater... dark chambers... fear and often panic... One day, while getting a detailed description... I asked the patient what was the size of her head in relation to her shoulders. When the answer was that the head was much bigger than the shoulders it dawned on me that I must be listening to a prenatal recall'.[138]

And therapists believe they can go back even further in their search for incidents of abuse.

'Repressed memories are a major cause of suffering for my clients,' said one Californian hypnotherapist. 'I see it every day, in about 90 per cent of the people who come to see me. There are quite a few physical symptoms that result from trauma in a past life. I really believe that all illness is psychosomatic. For instance, asthma may be the result of smoke inhalation in another life. An allergy to wheat may stem from a rape in a wheat field, or arthritis from being stretched on a rack during the Inquisition'.[139]

In a survey of more than 860 psychotherapists in 1994, a researcher found that 54 per cent believed that hypnosis enables the recovery of memories that date from birth and, astonishingly, 28 per cent agreed that hypnosis enables people to recover accurate memories of past lives. Just to be absolutely clear, this means that over a quarter of the psychotherapists surveyed hold the belief that people have lived a life before their current one, and that they can remember aspects of that life in some retrievable way.

I should put in a caveat here. The word 'psychotherapist' has no official standing or definition, and so it is perhaps too much to expect people who might have no professional training to be able to assess critically psychological theories based on rigorous research. Indeed, in view of this fact, it is perhaps predictable that people whose sole qualification for the job may be a certificate from a course taught by another 'recovered memory' therapist will come to believe in dreamwork, regression therapy, hypnosis, and reincarnation. Without those beliefs they won't have a job.

Many of the people quoted in Pendergrast's book are anonymous, including most of the therapists. But one therapist, Hamish Pitceathly, is happy to be named. He takes a robust view of the ability of people to remember back to their earliest days:

'Most people have some sort of scene such as this in their past. . . . One of the earliest scenes usually occurs between birth and six months old. The mother holds up the child by one leg and presses the child's head against her genitals. This can have a large number of effects. The most important is distrust, because after all, at that stage, the infant feels the mother is the entire source of sustenance. Suddenly the whole world gets turned upside down quite literally and metaphorically, and the damned thing is nearly asphyxiated—and in some cases *is* asphyxiated.'[140]

This man—he is quoted in Pendergrast's book saying many other equally strange things—is still listed on a hypnotherapists' register in the UK. He describes himself as Hamish Pitceathly MA, FSPCAH, Dip. PCAH, FHRS, FHS, MCPS(Acc), MNCH(Acc), NAHH(Reg), and the only one of these qualifications I can find spelled out on the internet is the Fellowship of the Society for Primary Cause Analysis by Hypnosis. The others appear to be equally obscure.

What is actually happening when people are regressed, according to one psychologist, Michael Nash, is that ' "age regression" is simply role playing in which an adult performs as she thinks a child would'. He points out that it is possible to obtain equally realistic and vivid accounts under hypnosis if patients are asked to project themselves *forward* instead of backward in time, to the age of seventy or eighty years, but presumably without any dramatic forecasts of what life will be like in fifty years' time.

One in four therapists believe in past-life regression, using hypnosis to send the patient back to a time before she was born, to a previous life, and they have taken heart from accounts of dramatic cases in which hypnosis subjects have revealed details of earlier lives which they could not have known and which have later been verified by more research. The most impressive evidence came in tapes recorded by a Cardiff hypnotherapist, of sessions with a Welsh housewife called Jane Evans. She had apparently led six previous lives, each of which was recalled in vivid and convincing detail.

I will describe just one of them, although the same points can be made about all. Evans claimed to have lived in the past as Livonia, wife of Titus, tutor to the son of the Roman governor of Britain. In a documentary film about her memories as recovered under hypnosis, viewers were treated to a realistic account on tape of a rebellion in York, where she lived, which led her and her family to flee to St Albans. She also gave a chilling account of the massacre of Jews in York in 1190. Yet when unhypnotized she claimed that she knew nothing of these stories, was unfamiliar with the Latin names, and had not heard of the massacre of Jews at York, which was a true event.

For some time, Jane Evans's previous lives were highly publicized in a book and in television documentaries and no one was able to show that she had taken the stories from some existing source. In fact, when Professor Brian Hartly, an authority on Roman Britain, heard the tapes he said: 'She knew some quite remarkable historical facts, and numerous published works would have to be consulted if anyone tried to prepare the outline of such a story.'

Hartly was right. Melvin Harris, an intrepid investigator of dubious claims, was convinced that Jane Evans had not lived six previous lives. On the other hand she didn't seem to have had the background or education to have consulted 'numerous published works', to put her story together. In the days before the internet, Harris faced an uphill task to find what he suspected was at the root of her remarkable stories—a novel about Roman Britain that she might have read years before and remembered in detail.

In fact, he eventually found the book, called *The Living Wood*, written in 1947 by Louis de Wohl, which contained characters, events, and locations identical with the stories Jane Evans told.

Evans had experienced a process called 'cryptomnesia' where some-one reads a book or poem or other text, forgets it, and then reproduces it thinking that it is their own work. The cryptomnesiac is often accused of plagiarism, although there can be a genuine lack of aware-ness of having ever seen the material before. So Hartly's statement that Jane Evans's story used facts from 'numerous published works' was correct. It's just that the numerous published works were consulted not by Evans but by the novelist Louis de Wohl.[141]

In Jane Evans's case, the hypnosis had no significant therapeutic purpose, but usually therapists use hypnotism to understand and then deal with some psychological problem. If the hypnotherapy reveals that the problem originated in a previous life, this is not much stranger than that the patient is remembering something from her first days of life. A past-life regression therapist described how he had successfully diagnosed one woman's problem: 'A 34-year-old woman came to me with a myriad of problems. The first thing she mentioned was this phobia she had about her husband's hands. Her marriage wasn't going well, but every time she tried to leave him, she kept getting this vision of his hands, and it stopped her somehow. So we went to regression. In a past life, she had also been married to this same man.'[142]

What's more—and here was the cause of the problem—her previ-ous-life husband had strangled her to death. Even worse, when she was regressed to another previous life, it turned out that, yet again, that same pesky man had been her husband, and this time he blinded her. The therapist recommended that when she went home she didn't spill all the beans at once about her husband's crimes in his previous lives. (Perhaps to prevent him strangling or blinding her yet again.)

A cooler-headed and less gullible analysis of hypnosis and what it can and can't achieve was published in 1997 by John Kihlstrom, a professor of psychology at the University of California at Berkeley. For a start, his definition is down-to-earth, and depends on no mysterious forces, swinging watches, or staring eyes.

'Hypnosis', Kihlstrom writes, 'is a social interaction in which one person, called the subject, acts on suggestions from another person, called the hypnotist, for imaginative experiences involving alterations in cognition and voluntary action. Among those individuals who are most highly hypnotizable, these alterations in consciousness are associated with subjective conviction bordering on delusion, and an experience of involuntariness bordering on compulsion.'[143]

Kihlstrom then goes through the research that has been done on various aspects of hypnotism, comparing groups who have been hypnotized and those who haven't as they carry out the same tasks. For the purposes of understanding hypnosis as used to recover memories of childhood, there are several well-argued conclusions:

Hypnosis cannot enhance performance, it can only produce the subjective impression that performance has improved without actual measurable improvement.

When used to test memory, hypnosis produces no enhancement of memory, but does increase the number of false recollections. This is clearly counterproductive in a therapeutic situation.

Age-regression under hypnosis is merely imaginative reconstruction. No one regressed to an earlier age shows physical or developmental signs that are characteristic of the younger age. Nor do they show evidence of being unable to access memories or skills that were acquired later. Again, subjectively they may feel like children and behave like children, but they are indistinguishable from unhypnotized people asked to imagine what it would have been like to be a young child.

Kihlstrom's conclusions, published more than ten years ago in one of the world's most prestigious science journals, the *Philosophical Transactions of the Royal Society of London*, might just as well have been published in the Basque language for all the impact they have had on the practice of psychotherapists involved in the recovered memory movement.

'These contemporary clinical practitioners,' Kihlstrom says of modern hypnotherapists, 'like their 19th-century Viennese forebears, rarely are able to obtain independent, objective corroboration of their patients' reports (or, for that matter, rarely even bother to seek it), and...uncorroborated memory reports are useless as scientific or clinical evidence about the historical past. In fact, there is almost no evidence supporting either the validity of the trauma-memory argument or the efficacy of recovered memory therapy. In the absence of objective corroboration from a representative series of cases, the use of hypnosis in recovered memory therapy lacks any scientific basis.'[144]

Kihlstrom sums up as follows:

First, the ability of young children to encode permanent memories of experiences is extremely limited, and there is no reason to think that hypnosis or anything else can overcome infantile and childhood amnesia. Second, hypnosis is first and foremost a state of believed-in imaginings: in the absence of independent corroboration, there is no reason to think that any hypnotically refreshed recollection is an accurate representation of the historical past, and, in fact, every reason to doubt it.[145]

10

· · · · · · · ·

Abuse of Truth

Some of the researchers who have contributed significantly to the understanding of childhood memory have undergone a change in their attitude to the topic as a result of that research. Gail Goodman, for example, of the University of California at Davis, started investigating the accuracy of childhood memory at the time of a number of high-profile allegations of child sexual abuse in day-care centres. Her motivation was a concern for the victims of child abuse and a starting point of 'believing the children'.

I wasn't fully on the false memory bandwagon like a lot of people [she told me], because I also very much wanted to be able to emphasize the limits of that work and that a lot of the reports are most likely true and there was a danger of overreacting either way... I think the McMartin case was certainly one of the big ones, and it took me a while to realize that maybe these *were* false reports. I think my tendency at first was to be concerned that these could be true but then as I learnt more about the case and actually got to see the interviews it was clear that the interviews had gone beyond anything I had studied.

But Goodman is still very concerned about the prevalence of child abuse, and in talking to her I always had a sense that she worried that some of the discoveries of memory researchers might prevent the detection and prosecution of genuine abusers. Because of this concern, she is also unwilling to believe some of the strong versions of false memory theory held by people like Elizabeth Loftus and Richard McNally. Her starting point is to accept that children's accounts can sometimes be false but always to bear in mind that much of what they say is essentially accurate:

When we were first doing our work on children's memory suggestibility, we assumed, as others did, that children would be highly suggestible and if you asked them questions like 'He hit you, didn't he?' and 'How many times did she kiss you?', that kids would go with that and make false statements. And we found actually that a lot of the kids were surprisingly resistant to those suggestions as long as they were four or five or older—three-year-olds still could be resistant but not as much so—but we were also interviewing them in a fairly neutral context. There's now been studies showing that if you are more aggressive you can get more false reports, but at that point, in the early 1980s or so, there were experts going into court saying, 'Well, if you ask a child those questions, of course they would make false reports', and so we were surprised that in fact they weren't doing that.

Part of the bedevilment of this field is the extreme positions promoted in courtrooms, on the basis of—but not necessarily by—expert witness testimony. One lawyer who spends her time prosecuting child sex abusers wrote an article for lawyers in which she said: 'Victim attorneys who pursue an aggressive strategy anticipating the false memory defense in abuse cases will help to level the litigation playing field.'[146] This rather suggests that the author thinks there is something unfair about the scientific evidence produced by researchers who can find no evidence for memory repression and who can demonstrate how false memories can be implanted. Aggressive strategies by *both* sides can lead to tragic injustices—sexual abusers remaining at large, certainly, but also entirely innocent parents or teachers having their lives destroyed by imprisonment and stigmatization. And, of course, the children at the centre of the miscarriages of justice usually have their immature emotional lives sullied by constant questioning and discussion of often bizarre forms of sexual abuse.

Underlying the difficulties of presenting some of this research in a dispassionate way is the uniquely awful nature of child sexual abuse. To defend an accused person in any criminal proceedings can always be interpreted as condoning the crime, even—in these repressed memory cases—before it has been established that a crime has been committed.

But there is a danger too in the blanket use of the phrase 'child sexual abuse'. Any sexual activity between a child and an adult is wrong, but even the words 'child' and 'adult' overlap in the teen years, depending on where you live. In New Hampshire, in the United States, it is legal for a girl of thirteen to marry. But in many of the cases we have been talking about, the occurrence of such abuse is treated as by definition traumatic when it may not have been. This is not relevant to the status of the abuse as a crime, but in two particular areas it is highly relevant—repression which is said to be due to the rather specialized issue of 'betrayal trauma', and the much broader issue of repression that occurs because the abuse itself—rather than any betrayal—is believed to be traumatic.

Jennifer Freyd's betrayal trauma theory can only be an adequate explanation of memory repression if abuse by a caretaker is perceived as traumatic by the child at the time. If a child suffers abuse that is seen as life-threatening or unpleasant or painful or disgusting, Freyd's theory suggests that the effect on the child will be even worse if, in addition, the abuse has been carried out by someone the child trusts and even loves. But for betrayal trauma to be a viable explanation, that trauma has to be there for the abuse to be recognized as betrayal. To make clear what I mean, a child will not feel betrayed if an adult does something to her that does not seem traumatic.

With the broader explanation offered for the repression of childhood abuse memories, it is the trauma itself that is said to be so intense and disruptive that it overwhelms the normal mechanisms of memory and leads to failure and hence to repression. But what if an episode of abuse is not traumatic? Is this even possible? As Gail Goodman's work quoted earlier shows, a stressful and undoubtedly traumatic medical procedure carried out at the age of two or earlier left no memory at all in the children she studied. And such memories, of medical proced-ures or sexual abuse, are not retrievable under any circumstances. With older children, certain forms of sexual abuse—fondling, kissing, etc.— might hardly be recognized as anything unusual and certainly not experienced as severe, life-threatening trauma of the sort that leads to PTSD for example.

In Susan Clancy and Richard McNally's work with people who claimed to have recovered memories of child abuse, the actual abuse was not always seen as traumatic. Sometimes it was the benign nature of the abuse that meant that they could not recall it, not because it had been repressed but because it wasn't significant enough to remember, as McNally explained to me:

What we found is that the typical recovered memory subject that we have seen is someone who *was* sexually abused, no question about it, if you look at it from a legal perspective, they're a sexual abuse victim when they were seven, eight, nine years old, but the abuse was not violent, there was no threat, they knew the person, an uncle, a stepfather, a grandfather who had a nice relationship with the kid, and the kid didn't fully appreciate what the heck was going on. So for example you might have someone who was sitting on a grandfather's lap while they were watching television, and the grandfather's fondling. The kid might think this is a little weird—'he's hugging me, he's touching me here'—but it's quite ambiguous. It's not traumatic in the classic sense of trauma—suddenly overwhelmingly terrifying and perceived as life-threatening. Many of these subjects of ours said, 'Sometimes I knew it must have been wrong what he was doing', because sometimes the perpetrators would say, 'Sssh, don't tell anybody about this, this is our little secret, our little game' and the kid would say, 'This is weird, why would he say "don't tell anybody"?'

Clarissa Dickson Wright, an English celebrity chef, has described such an incident of child sexual abuse in her autobiography. It took place on a sea voyage with her family when she was a small child:

One event on this trip stuck in my mind: there was a steward who in exchange for comics lured me into a bathroom and got me to wank him off. I found this fascinating, the growth of his penis, the velvet feel of it and the subsequent detumescence. I persuaded a little friend to come and share the experience and she told her mother who I heard created stink. My mother made no fuss to me so I suffered no trauma, the man was taken off the ship and my mother gently explained that some things were only for grown-ups. I was therefore unharmed by the experience and the fascination remains with me to this day. Years later when I was in treatment a counselor had the screaming abdabs at this story, appalled at such child abuse. No doubt it was but my mother's handling of it left me with no scars. Over the years I have met many people who suffered sexual abuse as children and one of the most consistent problems is the shame they feel, largely as a result of the reaction of discovering adults which makes the child think they were to blame in some way, so I have much cause to be grateful to my mother.[147]

Of course, there are many reading that story who would deny that Dickson Wright 'suffered no trauma' and was 'unharmed by the experience'. But without a reason for ignoring what she says about herself, it is a plausible conclusion.

So, as the child grows into an adult, this non-traumatic abuse, embedded perhaps along with all the other trivia of childhood, has no particular significance, and certainly is unlikely to cause the kind of post-trauma consequences that are suggested. But then, one day, as McNally pointed out, something happens to remind the adult, who looks at the abuse through adult eyes:

Suddenly, boom! she's reminded of it by something—a friend says, 'Oh, I think my daughter may be being abused', and then describes something quite similar—and all of a sudden, pow! the memory comes back. 'Oh my God, Jeez, that's what my grandfather did to me, wait a minute, we're related, I'm an incest survivor.' *Then* they are shocked and upset. And so the memory was not forgotten *because* it was a trauma, it was forgotten because it was *not* a trauma, just the opposite of what the repression thesis tells us to believe.

Jonathan Schooler calls this belated realization of the significance of child sexual abuse 'meta-awareness'. It's a process of looking back at the abuse through adult eyes and placing upon it an interpretation—a shocking and traumatic interpretation—which the child couldn't have comprehended. 'Later in life one revisits [the abuse],' Schooler said, 'and, through the lens of this new perspective, says "Oh, my God that was trauma—I was terribly upset about it at the time." '

One consequence of this, of course, is that the clients of therapists who believe that *any* child abuse causes serious long-term consequences may well end up being traumatized by 'meta-awareness' of an incident of non-traumatic abuse as a result of being told by the therapists how horrifying the events were. Seeing in a new light something they barely remember and never saw as harmful, such patients can begin to suffer psychological distress with a new belief of the significance of the event.

But defining some types of sexual abuse as less harmful than others, or even as harmless, is seen as dangerous by some psychologists, including Gail Goodman:

Even if a child doesn't recognize what's happening or doesn't see it as wrong, there still can be effects of it that we would categorize as harm, and then again there could be further harm, further traumatizing later. Some young kids, for instance, can become very sexualized by sexual abuse and may not see it as wrong, but is that a harm? There isn't really a great definition of trauma.

Scientists in this sensitive field may discuss in private such contentious issues as whether all child sexual abuse is harmful, but woe betide anyone who publishes research papers exploring this issue. One group of scientists tried to settle the issue with an academic research project and ended up being condemned by the US Congress.

In July 1998 an article by Bruce Rind, Philip Tromovich, and Robert Bauserman was published in the *Psychological Bulletin*. Its title was 'A Meta-Analytic Examination of Assumed Properties of Child Sexual Abuse Using College Samples'.[148] Scientific papers generally start with an abstract, a summary of the paper and its conclusions, and it's worth printing this one in full, in the light of the fuss the article caused:

Many lay persons and professionals believe that child sexual abuse (CSA) causes intense harm, regardless of gender, pervasively in the general population. The authors examined this belief by reviewing 59 studies based on college samples. Meta-analyses revealed that students with CSA were, on average, slightly less well adjusted than controls. However, this poorer adjustment could not be attributed to CSA because family environment (FE) was consistently confounded with CSA, FE explained considerably more adjustment variance than CSA, and CSA-adjustment relations generally became nonsignificant when studies controlled for FE. Self-reported reactions to and effects from CSA indicated that negative effects were neither pervasive nor typically intense, and that men reacted much less negatively than women. The college data were completely consistent with data from national samples. Basic beliefs about CSA in the general population were not supported.[149]

Although there's a certain amount of scientific language in this paragraph, it is not difficult to extract the significant points. The researchers pooled the results ('meta-analysis') of fifty-nine studies of college students in which other researchers compared two groups of students, those who had been sexually abused in childhood and those who had not, and tried to discover if, among those students who had been abused, effects of the abuse were discernible in later life. They found only one significant difference in the two groups—the abused group were, 'on average, slightly less well-adjusted'. Although in such research, the scientists try to compare like with like, there can inevitably be differences between the two groups other than the specific factor you are looking at. In this case, it turned out that the child sexual abuse group also had differences in family background. What's more, many surveys show that the types of 'marker' assumed to be effects of child sexual abuse are also typical of people from a certain type of family environment, where they experience neglect, lack of support or bonding, large family size, and so on. The two factors—child sexual abuse and family environment—are said to be 'confounded' and a further set of analyses had to be carried out to ensure that the inference—child sexual abuse leads to being less well-adjusted—is correct. When these analyses were done, the authors say, there was no longer any indication that poor adjustment is a specific consequence of child sexual abuse.

Now, I have no idea whether or not this paper has drawn valid conclusions from the analysis. Some people whose judgement I trust have told me it does; others whose judgement I trust equally have told me that there are all sorts of reasons why the conclusions the authors suggest are not supported by the evidence.

One of the factors that can get in the way of an objective survey of data is the personal bias of the scientists. However much he or she tries to discount such a bias it is difficult to prove that such a bias did not influence this work. It's somewhat more difficult to imagine such a bias in the case of three scientists working together on a project, but even if they were not steeped in prejudice, Rind, Tromovitch, and Bauserman may have shared a viewpoint about child sexual abuse before the survey that was confirmed by its results.

What I do know is that with its 52 pages, 13 tables, and more than 150 references, and conclusions which read as measured and moderate, presumably refereed by the authors' peers, this is likely to be no worse and its conclusions no less justified, than many scientific papers which find their way into scientific journals and whose results are accepted by the scientific community.

This was not good enough, however, for Dr Laura Schlessinger, American radio show host, who came across the article nine months after it had been published to little or no public comment. It was quoted favourably by an American website which promoted the legitimization of sexual relations between adults and children.

Schlessinger, not a medical doctor but with a PhD in physiology, described the paper as 'junk science', a phrase that was used several times in the furore that followed. She also said that it was endorsing paedophilia. With 18 million listeners throughout the US, very few of whom are likely to have read the relevant issue of the *Psychological Bulletin*, the news about this 'paedophilia-promoting' article was taken at face value, and the topic was taken up by various Republican congressmen. One of them, Tom DeLay, expressed his 'outrage and disgust' at the American Psychological Association for publishing 'a study that advocates normalising pedophilia'. Needless to say, the article did nothing of the sort.

What followed was described by officials of the APA as 'the political storm of the century for the field of psychology'.

Once the bandwagon was rolling, it was impossible to stop. It also provided a useful stick for the Republicans to beat Democratic President Clinton with, after they had failed to remove him from office four months before for what was viewed by many as the sexual exploitation of a young intern by a powerful, married employer more than twice her age. This was hardly paedophilia, but one issue could be used to remind the public of the other.

Resolutions were passed in various State legislatures, 'rejecting the conclusions'. Other resolutions insisted that the APA disown its own publication and condemn its own peer review system. The Republican Party sent a message to all Republican members of Congress claiming that the APA protected minors from paedophiles.

Finally, in an unprecedented move, the House of Representatives passed a motion unanimously which said that Congress 'condemns and denounces all suggestions in the article "A Meta-Analytic Examination of Assumed Properties of Child Sexual Abuse Using College Samples" that indicate that sexual relationships between adults and "willing" children are less harmful than believed and might be positive for "willing" children'. The Resolution also said that Congress 'encourages competent investigations to continue to research the effects of child sexual abuse using the best methodology, so that the public, and public policymakers, may act upon accurate information,' although at no point during 'the political storm of the century' did any of the objectors offer any detailed analysis of the article to show that it was not competent, or did not use the best methodology.

It is of course possible, as with any complex study, to suggest ways in which the conclusions of the Rind *et al.* paper might go beyond the evidence. I asked Gail Goodman what she felt about the study and she pointed out that, however well scientists try to devise a watertight study of the effects of child sexual abuse, it is very difficult:

How do you do it? You wait to see who gets abused, you get to find out about it, and now you look at the effects. How do you do a good study that controls for all the family factors in all these abusive families? You screen as well as you can. A lot of these studies didn't make a very job good of all that. And then you also add in: what is childhood sexual abuse? Well, if I see a guy flashing over here and I'm three years old that's sex abuse; if my uncle gives me a sexualized kiss, French kisses me, and I'm young, that's sex abuse; if I'm raped at gunpoint by a guy who's just got out of prison and comes to my house, and beaten and almost killed that's sexual abuse and so it's a huge range.... So it's conceivable that for me seeing that guy flashing it's horrible, and somebody else being fondled maybe kind of enjoyed it and didn't see it as that bad...

Another study, carried out on 1,400 twins a year after the APA storm, was able to come up with a more nuanced account of the effects of child sexual abuse on adults' psychological health. Picking up on the observation in the Rind paper that family environment might have been a confounding cause of psychological illness, the new work managed to separate out that factor more effectively because of the fact that the twins had been brought up in the same family environment but only some of them had been sexually abused:

Risk for psychiatric disorders increased as a function of severity of abuse; forced sexual intercourse was the strongest predictor of a range of subsequent disorders including major depression, bulimia, panic disorder, drug dependence, alcohol dependence, and generalized anxiety disorder... Shared family environment, often chaotic, was an implausible explanation for the connection between abuse and psychopathology: the risk for psychopathology was much greater in the twin who had been sexually abused than in the twin who had not been sexually abused, even though both had been reared in the same family.'[150]

The difference in the two conclusions didn't necessarily mean that the first paper was 'junk science' and the second wasn't. Each paper had differences of analysis and data handling which meant that they couldn't always be compared exactly. In particular, the reason that the two studies found different results in some areas might be because the later paper measured *lifetime* prevalence of psychiatric disorders, and some of the research in the Rind meta-analysis looked at *current* levels of distress.

I've told this story at some length because it illustrates the regrettable but understandable fact that, however neutral scientists think they are in their pursuits, the data they produce can arouse strong feelings among people who may not have the skills to assess the data but certainly know and say when they do not like the conclusions.

As the acknowledged 'queen' of the field of false memory, Elizabeth Loftus has experienced her share of public hostility, partly because her research is interpreted as playing down the importance of child sexual abuse:

People sent angry emails to all of my departmental colleagues. One blanket email began: 'Shame on you that you work with a person like Dr Loftus.' My 'enemies' tried to get professional organizations to rescind their invitations to have me speak. At some universities, armed guards were provided to accompany me during invited speeches after the universities received calls threatening harm if the talks were not cancelled. People filed ethical complaints. People tried to drum up letter writing campaigns to the Chair of my Department, the President of the University, and the Governor of the State to get me in trouble. A seatmate on an airplane once swatted me with her newspaper when she learned who I was. That's when I learned about dirty fighting.[151]

Many of the attacks on Loftus were triggered by what she sees as her most significant contribution to the repressed memories debate: several years of detective work which destroyed one of the key pillars used by repressed memory advocates in a number of court cases.

To recap, the concept of repressed memory of child abuse came about for two reasons. First, a climate developed that child abuse was rampant in society but unacknowledged and therefore unpunished. In the view of the people who supported this idea, one sign of this was a huge amount of psychological illness for which no explanation could be given. These two factors could be made to fit neatly together—if the psychological illness was caused by child abuse, this demonstrated that the abuse was as widespread as the illness. Because most people who are depressed, or have panic attacks, or obsessive compulsive disorder, or eating disorders have no memory of being abused as children this must mean that they have repressed the memory.

As we've seen, thousands of accounts of child abuse have been given to therapists and relatives and in courtrooms many years after the alleged events took place and after long intervals during which the self-described victims have had no memory at all of the abuse. As we've also seen, psychologists like Loftus and McNally and Bruck and Ceci have come up with alternative explanations for these accounts that do not require the creation of a new phenomenon of repressed memory (although they have created another new concept, false memory).

A point made by those who doubt that repressed memories can exist is that if they were a naturally occurring human phenomenon there would be some evidence of them occurring before the 1980s. Two senior Harvard psychiatrists, Harrison Pope and James Hudson, have offered a $1,000 reward for a historical example of traumatic memory that was repressed by an otherwise healthy individual and then recovered. The $1,000 has not yet been awarded. One psychologist, Ross Cheit, at Brown University, has called this 'a stunt'. He organized a website that listed 101 cases up to 2003 of recovered memory of being a victim of child abuse which he says are corroborated by independent evidence of the original abuse and which were repressed for many years after the abuse until allegations were made as an adult.

It is certainly an impressive list. It shows a large number of examples of adults remembering childhood abuse and suing or helping to prosecute the abusers. No doubt, the increasing amount of publicity given to child abuse over the years has attracted the attention of victims who might otherwise have been happy to let sleeping dogs lie. It is also a fact that some cases, at least, on Cheit's website could not have been brought if the victim had remembered the abuse at some point between the crime and the court case. This is because of a particular exception in US law in some states, where what is called a 'Statute of Limitation' applies, preventing a criminal case being brought for an alleged crime that took place a long time ago. (The time limit varies in different legal systems and for different offences.)

'If you wanted to sue somebody claiming that they sexually molested you twenty-five years ago, you can't just say "I forgot and now I remember",' Elizabeth Loftus told me. 'You basically have to say "I repressed the memory, I was unable and unaware and could have no access to this memory, it was walled off from the rest of mental life, and only after I went into therapy and felt safe and was able to retrieve all this information did it come back and therefore I should be allowed to sue." Because this notion of repression, or the idea under some other name, was sold to various legislators, we find in many states in this country that people can sue other people if they claim they repressed their memory.'

Furthermore, as far as corroboration of the abuse is concerned, Cheit's 101 cases sometimes depend on an admission of guilt by the accused. But once allegations of abuse are made against someone in a legal setting, it sometimes happens that a case is settled out of court with such an admission being made. This is sometimes wrongly interpreted as admitting liability, but some innocent people do this to avoid being bankrupted by legal costs.

But even if one accepts all the corroboration of the *offences* in Cheit's 101 cases, the corroboration of the memory as *repressed* is much less clear-cut. Most of Cheit's examples are based on newspaper stories, not the most detailed or reliable of sources, and even when he gives extracts from court transcripts, the proof of repression is not very convincing.

Cheit's Case No. 10, for example, is a case of abuse which undoubtedly took place, but it is on shaky ground as an example of a repressed memory:

Herald v. Hood (Summit County, Ohio, jury verdict, 1992; affirmed 1995). Julie Herald sued her uncle, Dennis Hood, alleging sexual abuse from age 3 (in 1962) through 15. The memory returned while Herald was watching her 4-year-old daughter play with a friend. Herald presented a taped telephone conversation in which her uncle indicated that she 'had been the only one.' Two therapists also testified that at a meeting with Herald in their offices, he admitted sexually abusing her. She was awarded $150,000 in compensatory damages and $5 million in punitive damages. The Ohio Supreme Court recently upheld the decision. (Reginald Fields, 'Witness Says She Felt Confusion and Guilt; Memory of Sex Abuse Comes Back by Observing Daughter, Court is Told', Akron Beacon Journal, July 25, 1992: C1.)

Ninety per cent of this account deals with the fact that abuse which was remembered had actually occurred. The only indication that this is based on a repressed memory is in the bald statement: 'The memory returned while Herald was watching her 4-year-old daughter play with a friend.' Although Cheit offers links to court transcripts, these do not provide any evidence that the victim had never remembered the abuse until the point at which she claims she recovered it. Cheit even gets this part of the case wrong. According to the part of the court transcript Cheit supplies, the memory was not recovered when the victim watched her daughter playing with a friend, but when she was hugged by an adult visitor to her house. (A false memory of Cheit's perhaps.) And yet, the only evidence which would confirm the case for repression is some way of proving that the victim had never told anyone before. As we know from Jonathan Schooler's work, backed up by laboratory experiments by other psychologists, it is sometimes the case that people who say they have not thought about abuse until 'recovering' a memory of it can actually be shown to have spoken to others about it in the intervening period.

Where the 'repressed' memory arises during therapy, there is another reason to doubt its accuracy. In another piece of research, Jonathan Schooler and Elke Geraerts investigated three groups of people who claimed to have been abused as children. One group had never forgotten, one group said their memory came back outside therapy, and a third group said their memory came back during therapy. With each group, the researchers sought the help of the subjects to uncover corroborative evidence of the original abuse. They then compared the three groups to see whether there were differences between them in terms of the amount of corroboration. The results were highly significant—in the ordinary sense of the word as well as being valid statistically. The accounts of abuse given by the group of people who had recovered their memories of abuse during therapy were far less likely to be corroborated than the other two groups. 'Memories that are discovered in the context of therapy are highly suspect', says Jonathan Schooler.

There was a further interesting aspect of this work. 'People in therapy [who had 'recovered memories'] were much less surprised at the discovery of the "memory" than people out of therapy,' Schooler told me, 'because the therapist was telling them they had it.' These were clearly not therapy sessions where a therapist tries to help a client discover the cause of her problems. They were sessions where the therapist *knew* the cause of the client's problems and shared the task of finding the 'evidence'.

There are various attitudes one can take to this particular skirmish in the Memory Wars. They span a spectrum. At one end are those who believe that, on the whole, if someone says she has recently recovered a memory of childhood abuse, in or out of psychotherapy, with or without the retrieval methods favoured by some therapists, then she should be believed until there is strong evidence that the abuse did not occur. At the other end are those who do not believe that such a process as memory repression exists, and that, if adults 'remember' abuse that they think happened to them in childhood, never having thought about it before or mentioned it to anyone else, they should be disbelieved until there is strong evidence that the abuse took place. Somewhere in the middle are those who do not rule out the possibility that on rare occasions a memory can rise into consciousness from a repressed state.

For Elizabeth Loftus one cast-iron, authenticated, case would shake her position at the far end of the non-believing end of the spectrum, and so when she read about such a case in 1997, she decided to investigate.

For some years during the 1990s a psychiatrist called David Corwin had been presenting at conferences a case which appeared to demonstrate unequivocally a recovered memory of child abuse that had been repressed for the entire period between the abuse and its recovery eleven years later when the victim was seventeen. Loftus told me about the point at which she began to get worried about the case:

People were talking about it as the supposed new proof and then I began to see it used in court cases against other people, including a court case to support the claims of a man who was on trial in Rhode Island. The expert in that case used this case history to try to bolster the complainant's testimony in that case where I believed that man was innocent. I would ultimately see testimony where Corwin himself used the case history in a case that involved accusations of abuse and used it to try and bolster the claims of the accusers. So it was being used in court cases against other people.

Corwin's recounting of the story of this specific 'repressed memory' started with him being called into a custody dispute between the parents of a child he called 'Jane Doe'. At the time, Jane was six and living with her mother, and Jane's father claimed that the mother was sexually and physically abusing the child.

Corwin interviewed the child three times in 1984 on videotape, and during the third interview Jane said that her mother 'rubs her finger up my vagina' in the bathtub, that it happened 'more than twenty times . . . probably ninety-nine times'. Backed up by his belief that the mother was unstable and the father telling the truth, Corwin gave evidence that helped to give Jane's father and his new wife custody of the child, and denied visitation rights to Jane's mother.

Eleven years later, Corwin wondered if Jane would remember anything of the abuse and he asked if he could interview her again. According to Corwin, this interview proved that Jane had repressed the memory of the abuse for the entire time between the age of six and seventeen. He would demonstrate this at conferences by playing a video of the moment the memory reappeared:

DC: Okay. Do you remember anything about the concerns about possible sexual abuse?

JD: No. (Eye closure) I mean, I remember that was part of the accusation, but I don't remember anything—(inhales audibly and closes eyes) wait a minute, yeah, I do.

DC: What do you remember?

JD: (Pauses) Oh my gosh, that's really, (. . . Closes eyes and holds eyes) really weird. (Looks at foster mother) I accused her of taking pictures (starts to cry and foster mother puts hand on Jane's shoulder) of me and my brother and selling them and I accused her of—when she was bathing me or whatever, hurting me, and that's—

DC: As you're saying that to me, you remember having said those things or you remember having experienced those things?

JD: I remember saying about the pictures, I remember it happening, that she hurt me.

DC: Hurt you, where? How?

JD: She hurt me. She—You see. I don't know if it was an intentional hurt—she was bathing me, and I only remember one instance, and she hurt me, she put her fingers too far where she shouldn't have, and she hurt me. But I don't know if it was intentional, or if it was just accidental.

DC: Can you be more specific because I—?

JD: I know what was said on the tape. On the tape it was said that she put her fingers in my vagina. And she hurt me.

DC: Okay. Is that what you recall or—

JD: That's what I recall. I recall saying it, and I recall it happening.

DC: You recall it happening?

JD: I recall. I didn't—that's the first time I've remembered that since saying that when I was 6 years old, but I remember.[152]

Loftus had various reasons for doubting Corwin's approach. She had seen him support allegations of child abuse against a female psychiatrist in a case where she, Loftus, believed there was no credible evidence that the abuse had happened. She was also troubled by the emphasis placed by Corwin on the instability of Jane Doe's mother, the uprightness of her father, the mother's troubled childhood, and so on. She wondered if there might be other explanations for the various accounts captured on video:

I thought two other things were possible. One is that the events never happened and they were suggested to Jane when she was a child. Another possibility is: maybe something happened but it was not really repressed and that she may have been thinking about it, discussing it, working on the material throughout the years until later on as an older adult she remembered it.

Adopting a new guise as the Miss Marple of the Memory Wars, Loftus decided to see if she could get to the bottom of what was really going on. Although Corwin only spoke about 'Jane Doe' in his paper and at meetings where he presented the story, there were certain clues in the videotapes themselves which Loftus saw. At one point he let slip the girl's real first name, Nicole, and the town where she had lived. He also mentioned that the child had been at the centre of a custody hearing.

Using this information, Loftus and a colleague found the court records, but these had preserved anonymity by removing surnames from the documents. Nevertheless, armed with the first names and the first letter of the surname, along with a further piece of information, that Nicole's father had died in 1994, she managed to find the details of the legal battles between Nicole's parents.

'I found that they had been in this acrimonious custody and support divorce battle for at least five years,' Loftus told me, 'and there was lots and lots of papers and information showing that it was a very vicious custody battle, and documents in that divorce file convinced me that this mother had been railroaded.'

When the mother was approached to see if she would agree to meet Loftus, she'd just sobbed and sobbed and said, 'I never thought this day would come.'

'I began to almost think of myself as a little Innocence Project,* ' Loftus said, 'because there is a falsely accused mother out there and I had to go and meet her. In any event what was unique about this case was the videotaping. A lot of therapists tell you a little story about their cases or some composite of their cases or some disguised version of their cases but there's no videotape.'

Next, Loftus and her colleague found Nicole's former stepmother working in a grocery store. She supplied one piece of the puzzle:

She told us about the months of efforts to wrest custody of Jane from her biological mother, and proudly admitted, 'That's how we finally got her—the sexual angle.' I came to believe that Jane's 'memories' which over the years had come to find a central place in Jane Doe's life, may have been powerful to her, but seemed to us to be probably false.[153]

Loftus spent months gathering material from the various people involved in the case and came to the following conclusion:

What I learned from this investigation is that there is very good reason to doubt that any abuse happened at all by this mother and there is very good reason to doubt the claim that Jane didn't remember or didn't discuss this supposed abuse throughout her life. In fact, I found evidence that is highly consistent with the idea that this information was suggested to Jane when she was a young child, perhaps ages four or five, before she was six. I found ample evidence from many individuals that she talked about the supposed abuse throughout her life before she went into her second interview with

* The Innocence Project at Yeshiva University in New York was set up as a training project for law students to reinvestigate past criminal convictions. It has been very successful in freeing falsely convicted victims of miscarriages of justice.

Dr Corwin. I can't prove that her mother did not abuse her but I certainly think that this work raises serious doubts about the allegations. This is not an ironclad proven case of repressed memory by any stretch of the imagination. It shouldn't be used that way, and it certainly shouldn't be used to advance any kind of scientific idea or in the persecution and prosecution of other innocent people.

Loftus was at the University of Washington at the time, and as she and her colleague were preparing to publish an account of their discoveries—in which Nicole's anonymity was preserved—the university administration dropped a bombshell. They phoned to say that someone would be coming to her office in fifteen minutes to take all her files related to the case. Nicole had complained that Loftus had invaded her privacy by investigating the case and the university had decided to take her complaint seriously.

After twenty-one months of investigation, Loftus was cleared of any misconduct, but matters didn't stop there. Dismayed by the attitude of the University of Washington, she had decided to accept an attractive job offer from the University of California at Irvine. Then Nicole filed a lawsuit under her own name, Nicole Taus, alleging invasion of privacy and defamation. Eventually in 2007 all but one of the counts were dismissed by the California Supreme Court, but one count, a claim of misrepresentation by Loftus, was allowed to proceed to trial. Later that year, that count was settled out of court between Loftus and Taus without Loftus admitting the misrepresentation. Taus, meanwhile, was ordered by the judge to pay the fees and costs incurred by others whom she also sued, to the tune of $240,000.

Loftus sees the whole process as an attack on freedom of speech. She argues that while some people really were abused and should have the right to speak about it and should obviously be believed, the problem is that there is a camp of therapists who appear not to apply critical reasoning to any claim of abuse which a patient makes, rather than starting from the standpoint that it is highly unlikely that 100% of allegations are true. It's not necessary to believe every claim in order to believe some of them, and moreover uncritical acceptance of dubious claims runs the risk of trivializing the real ones.

For Jennifer Freyd at the University of Oregon, just as important as academic freedom—sometimes more important—is that society is facing a widespread problem of child sexual abuse and should be doing more about it. She believes the scientists who are consuming resources to demonstrate false memory and refute memory repression could be devoting more attention to the bigger problem:

I have a hunch we could stop [child sexual abuse] in a reasonable amount of human time if we put the resources and made it a project, just as we have stopped other public health problems. We've eradicated diseases, we've made big impact on behaviours that are recognized to be harmful, and I think that if there was the kind of political and societal will that's being expressed about doing something about the obesity problem we could reduce childhood abuse to a relatively small problem compared with what it is now. It really worries me that by focusing so much attention, I would say disproportionately, on the possibility of a false allegation, that we have made it more difficult for people to get the support to talk in situations where it could stop something from occurring, and it has distracted our attention from what I see as this really big problem in our society.

Jennifer Freyd is a mother herself, and her undoubted worries about child sexual abuse and its effects may spring partly from that. But there is another possible reason for her concern. She has admitted publicly that she was herself a victim of child sexual abuse and, in an astonishing twist, spends some of her time attacking an organization set up by her parents to defend people from false accusations of child sexual abuse.

11

························

Freyds and Feuds

It can be a dispiriting activity to read the views of some of the practitioners of recovered memory therapy. These people appear to be living in a universe where logic and inference and deduction and evidence are fluid concepts, changeable at will to fit one's beliefs. Take a consultant psychiatrist, 'Delia Wadsworth', interviewed by Mark Pendergrast in his book *Victims of Memory*. Pendergrast anonymizes most of his interviewees, and perhaps this particular woman should be grateful to him for that. Her interview is riddled with the kind of illogicalities that one might expect a medical education to iron out. I thought of one of her more outrageous statements as I attended the 2007 Annual General Meeting of the British False Memory Society, in London. This was a gathering of a hundred or so members of the society, most of them parents who had been accused of abuse—they would say falsely—by one of their children.

Now, clearly it is possible that some of the members of this society, and its American counterpart, the False Memory Syndrome Foundation, are actually child abusers, people who have decided to live a lie for the sake of their reputations. Having denied to friends and families that the abuse they are charged with took place, it can be seen as a further testament to their integrity and the rectitude of their case that they then join such a society to campaign for the rights of themselves and other similarly accused parents. While it is conceivable that these societies have a few genuine abusers in their ranks, it is a big stretch to go from there to say that all of them are really child abusers, banding together in a self-protective huddle.

Pendergrast's interviewee 'Dr Wadsworth' says the following:

I believe that victims suppress the memory in a dissociative process. I think that perpetrators may also suppress it. They don't remember it either. . . . Have you ever been to meetings of the False Memory Society with these accused parents? The pain there is palpable, isn't it? There must be an element of suppression there. It's the only thing that makes sense to me. Otherwise, why would they appear to be in such bewildered pain?[154]

I truly don't understand what this garbled thinking actually means. It seems to me that the one emotion you would actually *expect* from people falsely accused of sexually abusing their own children would be 'bewildered pain'. To take that in some convoluted way as evidence that they had really committed the abuse they were accused of makes no sense at all. Presumably the logic is that (a) the parents genuinely believe they did not abuse their children, and (b) they have entirely repressed the memory of the dreadful events themselves. Setting aside that the evidence for total repression of being abused is thin to non-existent, Wadsworth's theory involves a whole new explanation.

For the argument usually offered for repression is that it is the intensity of the trauma that causes it in the victim. For *perpetrators* to repress the memory of their crimes would either require the *commission* of the abuse to be traumatic, which seems unlikely, or for there to be a separate type of memory repression which occurs when criminals don't want to face up to their crimes. No one, as far as I know, has produced evidence for such a phenomenon. One case that perhaps would come into the same category, at a stretch, involves the assassin of Robert Kennedy, Sirhan Sirhan, who claimed initially to have blocked out all memory of the crime. This later was shown to be a ploy to escape or minimize his punishment.[155]

Certainly, at the FMS meeting there was a lot of 'bewildered pain'. As I went into the conference centre in the heart of London and mingled with BFMS trustees and staff and sixty or so members, one of the trustees said to me 'Look at the grey faces—they're the ones who've only just been affected by this.' It didn't take long when talking to accused parents to get beneath the patina of apparent control and social ease to the depths of 'bewildered pain'.

In her Director's Report to the meeting, Madeline Greenhalgh gave a snapshot of a typical morning's phone calls and letters to the Society:

A phone call from a mother who had been accused by her forty-year-old son of abusing him when he was three. He was now involved with a hard line survivor's group.

A call from a barrister seeking advice on expert witnesses to help prepare the defence of someone who claims to be falsely accused on the basis of a 'recovered' memory.

A call from an accused parent seeking advice on where to find a solicitor.

A letter from a daughter trying to find out how to contact her father after ten years of estrangement, as a result of a 'memory' of abuse that came up during therapy.

A letter from a therapy patient inquiring about the scientific basis of age regression by hypnosis.

Enquiries from journalists following up abuse-related stories.

Over coffee I spoke to a Cornishman who had been part of a criminal trial in the West Country of England a few years beforehand, as a result of which he had spent three years in prison. The case was complex, messy, and involved undoubted serious child abuse by some members of a clearly dysfunctional family. This man had been drawn in to the family by marriage, and had therefore suffered from an assumption by the police that he must have been involved. As his story tumbled out it was full of accusations against the police, who, he said, had been determined to make a case and had just played around with the facts when it suited them in order to make the accusation stick. 'One of the charges said that I had committed offences at a date which was several years before I had even entered that family,' he said. 'When we pointed that out, the dates were changed.' When I asked for his email address he shook his head. 'I won't use a computer,' he said. 'You know what they can do these days with a computer. Planting stuff and that. Better not to.'

Another man I spoke to at lunch, an engineer whose daughter had accused him of abuse, spoke of the relief that he and his wife experienced when they came across the BFMS and its capable director, Madeline Greenhalgh. As he spoke about it he had to stop for a moment, overcome with emotion. You could see how cathartic it might be, having tried to cope with a situation you had never come across before and certainly never anticipated as applying to yourself, to discover that there were many hundreds of similar people who had been accused of child sexual abuse, and were now wrestling with disrupted relationships, suspicious family members, and even civil and criminal lawsuits.

When I raised the accusation, often put forward by those who believe in the repression of memory, that the BFMS was full of child abusers, the engineer said: 'Look, if you were really a child abuser, it's not the sort of thing you would want to tell *anyone* else about. Even when you're falsely accused you think twice about telling someone. But the outrage at the injustice of a false accusation is greater than that reluctance and it's *that* that drives you to seek the help and emotional support of a society like this.' This rang true with me. After all, if sex abusers all band together and pretend to be innocent, why aren't there established societies of murderers, burglars, and embezzlers doing the same thing?

A BFMS member has conveyed vividly what it feels like for accusations to arise out of the blue from a daughter who has come to believe that she was abused. Her daughter had been depressed, perhaps as a result of coping with the illness and death of a beloved grandmother. She was referred by her GP to a former travel agent who had taken up analytical psychotherapy:

After one session with the lady in question, [my daughter] thought she was the answer to all her prayers. Indeed every sentence started with 'so and so thinks, so and so says, so and so believes . . . ' We were asked first of all if we would mind paying the quite substantial bills for about a year (by which time of course she would be quite well again) and then if we would come and visit the therapist ourselves. It all seemed very positive and helpful—and we were completely taken in by this seemingly charming and helpful person. . . . She saw my husband and myself together and separately and asked all sorts of

174

extraordinary questions about our childhoods, our parents, our medical histories, and many things that seemed completely irrelevant. But, at that stage, we had absolutely no idea of what she had in mind for our family. It wasn't long before my daughter was given *The Courage to Heal* to read, asked to write down her dreams every night, and told to go to her doctor and look at her medical records particularly in relation to any childhood ailments. All this information was then 'analysed' by the therapist and gradually strange comments began filtering through . . . such as (from my daughter) 'the fact that I had my tonsils removed at the age of nine is a clear sign that I was being abused by someone'—and . . . 'apparently 99% of us are abused'. After a while, my daughter came home one afternoon and, quite calmly, told my poor husband that although she didn't remember it, according to her therapist she had been abused by him as a child. My husband was devastated and said that he didn't remember it either, and he assured her that it was not something he would have forgotten. It was all to no avail, and having been told that he was in denial, and that she would continue to see her therapist to put things right, she went away again, leaving behind two totally distraught parents.[156]

The British False Memory Society was set up a few years after an American equivalent, the False Memory Syndrome Foundation, was founded in 1992 as the wave of repressed memory cases in the United States grew. In the UK, as a result of the appearance of psychotherapy 'trainers' from the US eager to teach these new memory recovery techniques to British therapists, social workers, and police workers, there was a time lag. But the problem soon became as serious in the UK as in the US, helped by the publication of *The Courage to Heal* and similar 'self-help texts'. One estimate suggests that over 100,000 British families have been 'blown apart' by false allegations of sexual abuse.[157]

The formation of the FMSF in the United States angered supporters of the view that child sexual abuse is widespread, hidden, and tolerated:

Whether the False Memory Syndrome Foundation is a CIA front or a group of innocent parents tearfully wondering why their children don't love them anymore—or something in between—one thing seems certain: they [the Foundation] are a very, very bad idea. Because of the FMSF, much of the mainstream media in the United States is devoted to spreading the comfortable idea that abuse mostly doesn't happen, and that it's far more likely that a therapist is just messing with your head than that anything *bad* ever happened to you. Many a court case has been slammed shut because of misinformation from the FMSF. They taint the pool of information with discredited studies and misremembered 'facts,' and apparently, somehow, with sheer charisma.[158]

While the main value of the FMSF was for accused parents who had thought their problem might be unique, it also managed to make some therapists rethink matters:

In the fall of 1993, I attended my first local FMSF meeting. I wasn't sure what to expect. These were the accused, after all. I remembered all that I had learned about how all perpetrators are in denial. I expected a room full of defensive parents. What I found instead was a group of sad and shocked parents who asked the same question their daughters asked: 'How could she do this to me?' I had been so supportive of women and their repressed memories, but I had never once considered what that experience was like for the parents. Now I heard how absolutely ludicrous it sounded. One elderly couple introduced themselves, and the wife told me that their daughter had accused her husband of murdering three people. Another woman had been accused of being in a satanic cult that had used babies for sacrifices. This woman in a pink polyester suit was supposed to be a high priestess. The pain in these parents' faces was so obvious. And the unique thread was that their daughters had gone to therapy.

I didn't feel very proud of myself or my profession that day... I think that there is a point of no return with repressed memory therapy, where admitting what you have done to clients would be too terrible to ever face. Fortunately, I had not yet reached that point. Still, I left that meeting with a tremendous discomfort, realizing that I had clients who had cut off all relationship with parents who would have looked exactly like these people and would be in as much shock and disbelief. I felt like the sorcerer's apprentice.[159]

People who thought themselves victims of abuse also learnt that there was another side to the story:

My family had compiled articles and taped talk shows about something called false memory syndrome. I spent nine hours, all alone, reading this material and watching the tapes. I saw myself in them! I cried and cried, and I became very angry at Karen [a therapist]. I realized that my mind had been raped. I now feel that I was the perpetrator rather than the accuser. It is the shame I live with every day.[160]

Another 'survivor' was warned about the FMSF by the staff of the hospital where she was being treated:

They told me in the hospital to beware of this False Memory Syndrome Foundation, that it was a terrible cult full of pedophiles. I've met some parents in that Foundation now, and they're some of the nicest people you'll ever meet. One man at an FMS meeting was angry, though, and vented at me a bit. 'How could you let somebody put these lies in your head and believe them?' he demanded. I told him as respectfully as I could, 'It's not that I *let* anyone. That carries the connotation of giving them permission to screw up my head. I gave no one permission. I didn't even know they were doing it.'[161]

The founders of the FMSF, who themselves experienced the trauma of an adult child who claimed to have been abused, were an accused father and his wife. Their names—Peter and Pam Freyd.

Pam Freyd is an educational psychologist in her sixties and she works at the Foundation's headquarters in Philadelphia. Her daughter, Jennifer Freyd, inventor of 'betrayal trauma', who says she was abused by her father as a child, works three thousand miles away at the University of Oregon. For each of them the memory of psychological trauma, and specifically child sexual abuse, has been at the centre of their working lives. When I asked Pam Freyd if she wanted to discuss the circumstances of Jennifer's accusations with me, she said that she had agreed a 'truce' and that she would not talk about it unless Jennifer agreed.

When two people are at war, as Jennifer Freyd and her mother had been for the last fifteen years, you expect them to appear bruised. Pam Freyd had been sad and resigned when she spoke of the missing years since she last spoke to her daughter, but Jennifer Freyd when I met her seemed well-adjusted, alert, and focused on a field she had established as her own. When we talked briefly of the events that had led to the family break-up she just said, 'I'm a private person...'. The 'truce' I had heard about came as a surprise to Jennifer Freyd. In her view, she had always been reluctant to talk about her accusations of abuse and had only been compelled to when the storm of media coverage of false memories diverted attention away from what she saw as the major and unappreciated problem of child sexual abuse.

When discussing her work, Jennifer Freyd spoke passionately about this problem. She had recently co-written a leading article in *Science* that called for increased funding of interdisciplinary research into child abuse and the setting up of an Institute of Child Abuse and Interpersonal Violence. It was difficult not to see her interest in this field as a direct result of her parents denying her own alleged abuse when she had tried to attract attention to it. But Freyd disclaimed any simple connection between the two events:

The only thing I would want to say now is that the history of the timing has often been mischaracterized. Yes, I have my own private experiences but lots of people have their own private experiences. I was working in this area and then the false memory thing started off and then a lot of attention was diverted to me personally and to a set of issues, but it wasn't the issues I was interested in.

But Freyd did give an account of her own abuse to an Oregon newspaper in 1993. She said that she had been puzzled when she had suddenly become anxious about a forthcoming Christmas visit by her parents and didn't know why. The story continues, in the historic present:

Jennifer Freyd, 33, makes an appointment with a clinical psychologist, trying to make sense of her anxiety. During her second visit, the therapist asks if she was sexually abused as a child. Jennifer says no. But that day, the memories start. She had always felt uncomfortable with her dad's behavior: continual sexual talk, sitting in his robe so his genitals showed, suggesting she read 'Lolita' at age 9 or 10, showing her how grown-ups kiss for a part she has in a play.... [Later] Jennifer remembers that her father abused her.

Pamela Freyd wrote her side of the story in an article she published anonymously in which Jennifer's name was changed to 'Susan':

When she was 33, Susan had a revelation that she had been repeatedly sexually abused and raped for 13 years by her father. I am Susan's mother, and I have been trying to cope with that revelation.

In her article, Pam Freyd printed extracts from a series of emails she sent her daughter after the bombshell of 'Susan's' accusations:

178

Dear Susan,

My poor dear Susan. No one should have to have such secrets locked away.
How horrible. My poor child. You have memories of being abused starting
at age 3, of being forced into sexual intercourse between ages 14 and 16, of
being raped at age 16 a few days before you left for college. I struggle for
understanding. My heart weeps for you. I am so sorry for you. Alex ['Susan's'
father] has no memories of all this. I have no memories of all this. In our small
house, for so many years, how could all this have happened without my
awareness? . . . [162]

It is not my intention in reporting this train wreck of family relationships to come in between the two approaching locomotives. (The
Freyds have been called 'the most influentially dysfunctional family in
America', by someone called Stephen Fried, no relation.[163]) Neither
Jennifer nor Pam showed signs of severe psychological disorder when I
met them, but the argument raises a question which underlies much
controversial scientific research—as well as the nature of memory
itself—as to the role of subjective beliefs in attempting to arrive
at objective 'truths'. No wonder Jennifer Freyd is so passionate about
anything that might minimize the attention given to child sexual abuse
in our society. And no wonder Pamela Freyd and her husband felt
moved to do something for parents they saw as being in a similar
situation to their own with no one to turn to at the time.

Richard McNally described the clash of viewpoints to me as
a 'Rashomon' effect, where several people sincerely believe in contra-
dictory versions of events, but, in spite of his lack of belief in the
repressed memory phenomenon, he had only praise for Jennifer
Freyd's professional work. 'She's actually a very eminent cognitive
scientist,' he said to me. 'She was working in a different area, then
she moved into the area of memory and trauma and so on. She's
the real deal.'

And Jennifer Freyd herself brushed aside any suggestion that her
choice of research topic was conditioned by her personal experience.

One of the things that really strikes me is that I know a lot about rates of
trauma and I know that the rates of child sexual abuse are high, and with most

people their private lives are not open to public discussion. But I know in the field that many people have a history of trauma but the base rate is so high that it's not surprising. But it is interesting that most people jump to the conclusion that one is interested in it *because* of that, but if that were the case over half of Americans would be trauma psychologists.

As I walked back to my car from Freyd's office on the sunny, autumnal University of Oregon campus, through a throng of students inspecting racks of clothes labelled '$5 all items', or chewing on falafel and hummus wraps, I thought, if Jennifer Freyd was right about the high rate of prevalence of early sexual abuse in the population, far more of these apparently cheerful, well-adjusted students had been abused sexually as children than had ever been appreciated.

Robyn Fivush at Emory University is another researcher who feels uncomfortable providing data that may support the false memory hypothesis, even though, of course, she still publishes such results:

As adults, 25 per cent of women claim that they have had some kind of molestation or penetration by a family member or close family friend before they were sixteen that they did not want. So if that's your definition of sexual abuse you have about 25 per cent of adult women in this country who were sexually abused.... So I think it's prevalent, I think it happens to an awful lot of people, and I think most people never talk about it, but most people remember at least parts of it—at various points in their lives it may become more or less present. Don Read and Steve Lindsay have this wonderful study that shows that people will forget that they went to summer camp. So it's not that forgetting doesn't happen or at least it's so pushed up out of awareness that until you tune in you say 'Oh, my God, I'd forgotten all about that.' I don't think sexual abuse is like going to summer camp. Your life changes, you're living in a different place, you're a different age, you're living with different people, things get pushed to the back, you don't think about them. It doesn't mean that then when you remember it's a false memory, nor does it mean that it's completely accurate as you recall it. There's some middle ground here, that's what I'm trying to argue, so somebody who's watching Oprah and there's a show where somebody's talking about early childhood abuse and they say 'My God that's happened to me'—that's possible. That's cued, all memory is cued.

Fivush looked around the hotel courtyard in Atlanta where we were meeting, and pointed to some plants:

I'm going to remember these flowers because three years from now I'm going to be walking in a botanical garden somewhere and go 'Oh my God, that's just like the flowers...' and nobody will think there's anything bizarre about that and say 'you're creating that memory, that's a false memory...'.

By saying that all this is false memory creates a science of memory in which we have no accurate memories and that's distorting science. And I also believe, as a feminist knowing what the epidemiological proportions are, that we are doing women and society a disservice by yet again pushing violence against women and children aside and claiming it's not real.

But even if Fivush and Freyd and others are right about the prevalence of child sexual abuse, how much should that fact weight the scales against the needs of parents who say they have been falsely accused, or, indeed, the views of children of those parents who come to believe they have been unduly influenced by books like *The Courage to Heal* or enthusiastic therapists?

Robyn Fivush said that it is distorting science to say that 'we have no accurate memories'. I agree, but I'm not sure anyone is saying that. What many memory scientists *do* say is that some of our childhood memories are false memories, and that there is no guaranteed way to tell uncorroborated true memories from false.

Whatever the dangers of disbelieving children and adults who have been genuinely abused, it seems to me that, based on the kind of research I have reported in this book, we can no longer automatically 'believe the children', unless we are prepared to change the basis on which we believe *anyone's* narratives. Judging by the discoveries of Bruck and Ceci and others, what makes memories *seem* credible, particularly as we listen to them being retold on different occasions, can often be what actually marks them out as false:

An intriguing experiment carried out over fifty years ago suggests that not only can experts not judge true memories from false, we sometimes don't even recognize our own memories when we haven't thought about them for a long time:

Twelve subjects in group therapy were asked to recall childhood memories involving parents, siblings and sexual experiences. The Freudian therapist conducting the study was particularly interested in stories about rejecting fathers and flirtatious little girls. The memories were transcribed onto a pack of cards, shuffled, and presented to the subjects from three months to four years later. *None of the patients could identify all of their previously reported memories.* On average, they correctly recalled *half* of them.[164]

12

· · · · · · · · ·

Truth or Consequences

Whether a statement is true or not might seem to be a matter of some importance in our society. It is certainly important in politics and science and law and relationships—the cement that holds together modern societies. But when it comes down to individual childhood memories, it could be argued that their truth doesn't really matter. Whether a cow really did bite Madeline Eacott's sister or, as seems to be the case, Eacott hit her, is the stuff of family folklore, to be laughed about at Christmas get-togethers but of no real consequence.

The law has implicitly recognized the fallibility of human memory in the establishment of a statute of limitation in some legal procedures such that new accusations of an assault that took place decades ago cannot be made after a certain lapse of time. This is partly practical, of course. No one wants to waste time trying to find evidence and witnesses of events that took place in the distant past. But it is also a recognition that memory is fallible and the longer ago a memory refers to, the more likely it is that errors will creep in.

Nevertheless, it might seem obvious that truth is *desirable* in memory, at some level. It might seem perverse for someone to say that it doesn't matter if a memory is historically true or not, and yet some philosophers and psychologists say precisely that.

Donald Spence is a psychoanalyst whose book, *Narrative Truth and Historical Truth* (1982), put forward two views of memory as expressed in therapy, only one of which has much to do with whether the remembered events happened or not. Spence's 'historical truth' is what applies when a patient says 'X happened' and it actually did, at some place and time related by the patient. But he also writes of another form of truth, 'narrative truth', in which events which may never have happened are treated as true. He defines narrative truth as 'the criterion we use to decide when a certain experience has been captured to our satisfaction; it depends on continuity and closure and the extent to which the fit of the pieces takes on an aesthetic finality.' Another word he uses about narrative truth is 'conviction'.

This to me sounds dangerously like juries who decide that a witness is telling the truth because the details he gives are so convincing.

In psychotherapy, according to Spence, 'narrative truth has a special significance in its own right...making contact with the actual past may be of far less significance than creating a coherent and consistent account of a particular set of events.'[165]

I suspect that the psychotherapists Spence mixes with are highly trained, intelligent people who can easily handle the nuances of the differences between his 'historical' and his 'narrative' truths. Clearly, in psychoanalysis and some of the more dynamic psychotherapies, useful inferences can be made on the basis that what is important to the patient is revealed by his or her 'memories'. But to use the word 'truth' in this context just seems to me to muddy the waters and lead to some of the crack-brained ideas that infest the world of repressed memory therapists.

I think the fewer meanings a word has, the less opportunity there is for misunderstanding, and to talk of a memory as 'true' when the event being remembered didn't happen just seems perverse. The memory may be significant or revealing or symbolic of something true, but it isn't true if it is false.

Unfortunately, judging by Mark Pendergrast's interviews with therapists, the Spencian view of the flexibility of truth seems to have caught on in a big way. Here are comments from two of his interviewees:

In therapy, it doesn't matter whether it's verifiably true. One's aim is to get the patient better. I know when someone is telling the truth. I can just tell. It's experiential. I usually can't prove that they're telling the truth.[166]

We have to accept [the client's truth] whether it existed in objective reality or not. How is that truth playing out in the client's life? It's the *dynamic* past, not the *content* past, that is important. If somebody thinks they were orally penetrated at nine months old and objects were inserted in her vagina, then if they have a gag reflex and can't eat solid food and can't enjoy sexual contact with their partner, it seems reasonable to look back, accept the story, expand it, and explore with that person.[167]

The *un*reasonableness of such an approach becomes clear when it leads to innocent people being hauled up before the courts.

One accused father described being interviewed by the police on the basis of a litany of abuse contained in what would presumably be called a narrative memory by some therapists:

I had supposedly been abusing Emma from the age of three. When she was fifteen or so, I had somehow got her and a whole group of grammar school boys to perform sex acts on the stage of the school. Don't ask me how I got into the school. Then a year or so after that, I decided she had to become a prostitute, so I took her off and introduced her to men I knew who paid her for her services. Then, not content with that, I decided to form a satanic ring for ritual abuse. I got together a whole crowd of men, mostly civic leaders from Devon, a fire chief, a lot of the people who worked with me, about twenty men altogether. And don't forget the vicar and a doctor friend. So it was a nice little group we had going. We all dressed up in black robes with yellow sashes. There was apparently one other girl there, the daughter of a friend of mine. We tied her and Emma down on a big oval table in my office. It all happened at my office.[168]

To set against this 'narrative' truth was the historical truth that emerged from a medical examination, that Emma was a virgin.

I have written a lot about the misdeeds of therapists but I am happy to accept that standards of training have been tightened as a result of the scandals and tragedies arising from false accusations in the 1990s and that, it is to be hoped, there will be fewer cases that get through to the courts on the basis of uncorroborated and often implausible descriptions of abuse from decades beforehand. There is, however, a potentially worrying development in the UK. In October 2007 the government announced a new initiative in mental health such that 'everyone who needs it should have access to psychological therapy'.

The emphasis in current NHS guidelines for psychotherapy is on cognitive behavioural therapy, a reasonably evidence-based method, but the rapid need for the 3,500 new therapists required could lead to poorly or inappropriately trained people being let loose on the public, and 'recovering' more memories of abuse.

However, as we've seen, there are many people whose belief that they were abused did not start with a therapist at all, but with a single book, which is still freely available in the UK and the US, *The Courage to Heal*. It is offered as a guide for survivors of sexual abuse, and promoted by a range of self-help organizations for survivors, as well as in a report by the Church of England.[169]

Its pernicious effects lie not only in the way it sets out as if proven a theory for which there is little or no evidence. It goes further than that—it provides rules for how to detect signs of abuse in the reader, and case histories which provide examples of how to accuse the alleged abuser and break off contact with the family. This book has more unsupported statements per page than any 'serious' book I have ever read.

In a booklet published by the British False Memory Society, *Fractured Families*, many parents testify to the role *The Courage to Heal* played in their child's decision to make accusations of abuse. One extract shows how this works:

One morning in July 1997 at 7 a.m. the doorbell rang. My husband, John, answered the door to find two detectives. The senior one accused him of committing acts of sexual abuse. John was escorted to the local police station. Police officers searched our home, and further officers questioned Julie, our younger daughter, at her home. After being detained for about eight hours John was driven home by one of the officers and told not to be unduly worried about the allegations. The police soon realised that Susanne's 28 page statement contained many inaccuracies, and parts appear to have been copied from the book *The Courage to Heal*.[170]

Another accused parent wrote:

The Courage to Heal . . . clearly influenced some of the emails we had received e.g.: 'What went wrong in my childhood? Why do I have no real memories before the age of eight? Why has this period of my life been wiped out of my mind? Something so bad must have happened for this to have occurred.'

Compare this with the section in *The Courage to Heal* headed: 'But I don't have any memories' (p. 81) 'But I had to ask myself: "Why should I be feeling all of this? Why should I be feeling all this anxiety if something didn't happen." If the specifics are not available to you then go with what you've got.'[171]

'Going with what you've got' in making accusations of child sexual abuse when you have no memories of it, is one of the more irresponsible pieces of advice in the book.

The Courage to Heal has a sober and inspiring title, closely printed pages, a long list of 'thank-yous' and a nineteen-page bibliography. But the closer you look, the flakier it seems. The book neglects the possibility that there are other causes for mental illness than incest and abuse; nowhere in the book, let alone in the references, is there any mention of the scientific work on false memories, landmark cases such as McMartin, Wee Care, Ramona, or any of the dozens of cases of false accusations, abandoned or rejected lawsuits, recanting 'survivors' or therapists.

As it was published in 1988, and the version currently on sale seems to be the same text, it is perhaps not surprising that it doesn't include the groundbreaking researches and *causes célèbres* of the 1990s that would destroy its arguments. What *is* surprising is that its publishers, editors, agents, and authors see no need to withdraw the book and, if they want to republish, at least to have it updated by independent and better-informed researchers. I believe it is possible that, by ignoring changes in our knowledge of the field, the authors and publishers may mislead vulnerable people into believing things about the origins of their own distress that are just plain wrong. In June 2003, in the House of Commons, MP Claire Curtis-Thomas said: 'Many responsible psychiatrists and therapists regard it as one of the most dangerous self-help books ever written.'

In fact, the book has the effect of a virus. Propagated through therapists and survivors groups, its messages summarized, abbreviated, and transmitted as slogans—'If you think you were abused and your life shows the symptoms then you were'[172]—*The Courage to Heal* has acquired the status of a Bible for survivors, when it is no more than an outdated fundamentalist tract.

The book's viral nature was emphasized for me when I came across a page on the Carnegie-Mellon University website in the United States in 2007 which quoted almost verbatim the message of *The Courage to Heal*:

How can I know if I was sexually abused? If you remember being sexually violated as a child, trust your memories, even if what you're remembering seems too awful to be true. Children simply do not make these things up.... Whether or not you have specific memories, if you suspect that you were sexually abused, then you probably were. Often the first step in remembering involves having a hunch or a suspicion that some type of violation occurred. Pay attention to these feelings, for people who suspect that they were sexually abused generally discover that this has been the case."[173]

This page has now been removed, after the University's attention was drawn to the erroneous information on its site, but similar content still appeared on the website of the University of Illinois, as this book went to press.[174]

I have tried to show in this book how—as I believe—the only way to discover the truth of a statement, or at least to maximize the chances that a statement is true, is to use rigorous, well-executed, statistically valid scientific techniques. It seems to me that books like *The Courage to Heal*, and therapists who use recovered memory or age regression techniques, make statements that are every bit as testable as that 'the mass of the electron is 1/1836 of the mass of the proton', but some of the people behind these statements appear to ignore the explanation that is most likely to be true, the findings of well-conducted scientific research into the flexibility and fallibility of memory.

Having investigated the inchoate mass of inaccurate fragments that make up childhood memory, scientists have come to see that at no point in our lives can memory really be said to be 'accurate'. Accuracy is a concept that is only peripherally related to the function of memory in our lives, and to expect photographic precision and chronological accuracy is not only unreasonable, it is unnecessary. Why, indeed, would it be desirable for a sixty-year-old to be able to relate in graphic detail the daily events of his second or third year of life? Or even to be able to say in 2008 exactly what he or she was doing on 8 March 1992, unless there was some specific significance to the date?

A trivial visual impression that was linked to a crucial emotional experience might be far more important to remember than the details of a bland visit to the supermarket in our pushchair or a journey to school on the bus. A child's experience that might seem mundane to an adult could have been an occasion for the child to learn something for the first time that became part of that child's personality and behaviour in the future. The discovery that it was possible to feel certain and yet be wrong—my Pratt's/Spratt's incident—was an early event in my life that led to 'never assume' being one of my mottos, and yet it came out of a bus journey along Streatham High Road which I probably took hundred of times without being able to remember any of the others.

While science has devised better and better ways of getting at the 'truths' of human psychology, the popular idea of memory has stayed firmly stuck in the videotape model. And yet, our own introspection can show that this model is wrong. Some of the respondents to my informal email survey realized that memories that were firmly projected on to the virtual screens in their minds had never happened. And yet that realization was intellectual rather than immediate—the man who remembered his father playing with the family Labrador in rural England still has that memory even though he now knows that his father was in Egypt at the time, never to return.

Nevertheless we still question our children about their memories, in the courts as well as in the playroom, and treat their answers as if they were accurate. Or at least, the *starting point* for assessing their memories is to listen and believe, until we come up against some inconsistency. Then, perhaps, we begin to accept the fallibility of childhood memory.

Unfortunately, there is still a long way to go. Very young children are still questioned by police and social workers who believe in the 'videotape' theory and, furthermore, know little—or at least believe little—of the mass of data that has been gathered about the pernicious effect of suggestive questioning on shaping children's accounts of their experiences so that they say what the interviewer wants to hear. Furthermore, not only do children say what authority figures want them to—they can actually come to believe inaccuracies and falsehoods themselves. Science has shown that whatever the starting point for a childhood memory—accurate or inaccurate—things can only get worse as the child gets older. No memory ever got *more* accurate with time, and there is a mass of influences in our daily lives which make them *less* accurate—personal introspection, natural memory decay, conversations with others, desire to remember or forget, wish to be seen in a better light, misplaced associations with similar impressions, are just a few.

As someone who believes in the power of the scientific method to provide useful insights into how we and the rest of the universe work, I find it depressing to realize that many of the discoveries of well-conducted experiments in psychology have failed to penetrate to the general consciousness. And in spite of what some people believe, the scientific method can be applied as effectively to psychology as it can to physics or chemistry. It all depends on the scientific design. It's true that the human brain is far more complex than, say, molecules of sodium chloride in water. And, of course, the scientific method is a long way from telling us everything about the brain. But it can lead to discoveries with the same degree of certainty that applies to discoveries in the physical sciences (although in neither is it usually 100 per cent). When scientists at Harvard show that the emotions that accompany planted false 'memories' of witnessing a fight between your parents are indistinguishable from those that accompany true memories in people who have actually had the experience, that is a valid result which can be repeated by other scientists with other subjects. And when Jonathan Schooler demonstrates that some people who claim not to remember a particular event actually spoke to someone about that event some years before, he is making a scientific statement that is every bit as valid as a biochemist who says that the cysteine redox sensor in PKGIα enables oxidant-induced activation in living cells.

But scientific results don't get very readily into public consciousness. Compare McNally's results—repeatable by any other psychologist who follows the experimental protocol—with the sort of statement that is made every day to patients who question their therapists about the rationale for their approach. When a therapist said to a patient, 'The body never forgets,' and the patient said, 'Yes, but where is the scientific evidence for this?' she got this answer from the therapist, who became quite agitated: 'You must trust me.'[175] On the whole, scientists don't have to rely on our trust when they announce new results—anyone can read about and assess the reliability of their research.

Ever since Frederic Bartlett's pioneering work in the 1930s, scientists have been able to show that memory is a reconstructive—rather than a reproductive—process. Later research has shown a dozen different ways in which that reconstruction takes place. First, even the act of placing a remembered event in the right time-frame is a work of reconstruction. An individual memory has no time- or date-stamp. The only way we can tell *when* a remembered event occurred is by inference. Perhaps we know that it happened when we were at school or university, or in a house we didn't move to until after 1980, or that it happened while someone was alive who is now dead, or that it was on a Tuesday, but which Tuesday is impossible to fathom.

Then, when it comes to the details of a memory, we may not remember things either because we never took them in, or we took them in and didn't pay enough attention to them to consolidate them, or we took them in, consolidated them, and then they were changed as a result of recalling them in a context where other factors intervened to change them.

It has been science and scientists who have also demonstrated the extreme malleability of memory, in laboratory experiments that have created false memories which for the experimental subjects become every bit as real as their 'true' memories (which of course we now know are not necessarily true either). Further, this research has identified the factors outside the laboratory that reshape old memories and implant new ones, factors which are all too present in police stations, social workers' offices, and therapists' consulting rooms.

But no scientist has ever demonstrated that it is possible for thousands of people to undergo severely traumatic experiences as children, entirely forget those experiences for years or decades, and then recover them in vivid and circumstantial detail. Even the Jane Doe case that was put forward as a single verified example of true repressed memory—and quoted in trials to support the idea of recovered memories—was shown by Elizabeth Loftus to be nothing of the sort.

The situation is actually worse than this. In spite of efforts by some well-meaning memory scientists, there is no convincing theory for why such repression might take place. Or at least, such theories that exist are very difficult to substantiate. Not only has no reliable evidence of memory repression been produced—there is a large amount of evidence that traumatic events which occur at any age above two or three leave lasting and ineradicable memory traces. The difficulty for those who have suffered trauma is to forget, not to remember.

One unfortunate message that has come out of the last two decades of work in this area is that, in Lenin's phrase, 'a lie told often enough becomes the truth'. In the case of the repressed memory theory, the lie has not only become the truth in an intellectual sense, so that many uncommitted people come to believe that such a phenomenon exists. It also becomes a personal truth for the unfortunate people who are offered repressed memories as an explanation for their problems, by therapists and by books like *The Courage to Heal*. Worse still, it becomes a 'truth' that convicts parents, relatives, and teachers of crimes they didn't commit.

While 'the worst catastrophe to befall the mental health field since the lobotomy era' may be on the wane, the danger has not passed in spite of the case against recovered memories built up by two decades of scientific research.

The pace of scientific research is not always as quick as we would like. Theories have to be devised, experiments designed that are rigorous enough to ensure that whatever they show has a high probability of being true, funds have to be raised to pay for the research, subjects have to be recruited, and the requisite number of trials have to be carried out. Then, there is the statistical analysis of the results, the exposition of the conclusions, and the writing and publication of the scientific paper. All of this can take years. But the fact is, since the idea of repressed memory first reared its head in the late 1980s, the work has been done and the results published.

Unfortunately, there are people still in prison as a result of the mistaken beliefs that were promoted in courtroom arguments by zealous prosecutors in the course of the wave of recovered 'memories' of child abuse. Legal action was encouraged by therapists and social workers who believed in the existence of a phenomenon that, if it exists at all, relates to only a tiny proportion of cases of child sexual abuse. The reason for this is partly that, however slowly science works, it is the hare to the legal system's tortoise. First, it is not always easy to get recent scientific evidence put before the court, particularly in the light of the rule in some jurisdictions that scientists cannot be called as expert witnesses to talk about something which an ordinary juror would be expected to know already, and it is assumed that everyone already knows enough about the anatomy, physiology, cognitive science, and psychology of memory. Then there's the fact that once someone is convicted, it is only new evidence that can justify an appeal or the overturning of a verdict, and that doesn't mean new scientific theories, but merely some new fact that might come out about the 'crime'.

Perhaps one of the most *anti*-scientific aspects of this whole story, which is peripheral to the science of childhood memory but central to the difficulties of exploring and publishing some of the new results, is the contentious nature of the crime of child sexual abuse which is at the heart of the issue of recovered memory. Sincere and able scientists feel uncomfortable about discussing or supporting data which might be used to diminish the enormity of genuine child sexual abuse or suggest that its effects were not deeply traumatic to the later life of the child. We only have to consider the unanimous vote in the US Congress to condemn the *Psychological Bulletin* paper suggesting that child sexual abuse does not always cause intense harm on a pervasive basis. This condemnation was not based on any scientific assessment of the data but on general uneasiness that such a conclusion might be correct.

193

There is a final lesson in all this. While this book started with those memories that we all have of childhood, the results of research into childhood memory, 'repressed' memories, and false memories has shown that *all* memory, whatever age it is laid down or recalled, is unreliable. The effects that have been revealed of suggestive questioning, the role of authority figures, the misuses of hypnosis and guided imagery, the appeal to the value of remembered details in assessing whether a memory is accurate or not—all of these abuses and more operate at every level in life. Children may be more suggestible than adults but the evidence shows that we are *all* suggestible to some degree, particularly when put on the spot in a police station or courtroom. The Innocence Project at Yeshiva University in New York has shown the pernicious role of eyewitness testimony in 75 per cent of wrongful convictions. Somehow, hundreds of witnesses 'remembered' lies, or were persuaded to believe them. And any of us could be one of those witnesses.

This is not a reason for pessimism. Science has not only shown that memories are misleading. It has cast light on the true function of memory, where gist is as important as detail. What does it matter if the sabre-toothed tiger attacked you on a Tuesday or a Wednesday? Or if it came into the clearing from the north or the east? It *was* a sabre-toothed tiger, it *was* in that clearing, it nearly cost your life, and you won't go back there in a hurry.

But regardless of issues of the accuracy or inaccuracy of memory, it's clear that our memories are much more intertwined with our identities than had previously been thought. From the mother (usually) who discusses the day's events with her child, to the self-image we create as adults, we sculpt our memories to fit within the outline of who we are, or, just as often, who we would like to be. It wasn't the number of ducks on the pond or whether they were on the green grass or the green bank that mattered, but that I had a mother who recited the poem to me while I sat on her knee.

Appendix

A selection of childhood memories, gathered for this book, in order of the age at which the events occurred. Where no other details are given, the source is the author's own database.

1
Sunlight sparkling through leaves.

1.5
We went on the gold sands and I saw black cows leaving their cowpats along the way—under the cliff at Manorbier in Wales it was in fact—and then we were in a bit of a castle in Manorbier where friends lived and there was washing up and knives with ivory coloured handles and those rounded regular blades and the handles square and being told not to put them in too much hot water. And soapsuds. Being in some kind of pram and the yellow flowering branches over my head and it was warm and sunny and I was happy.

1.9
I remember the look on my mother's face as she grabbed me up and ran with me downstairs. I remember the house shaking and bottles falling around us.

1.9
Car stopped, getting out at odd grassy place, teddy or doll needs to get out too. Am with mother and grandmother, I think.

1.9
Waking up in cot, lying, seeing yellow curtains waving in breeze, somebody nearby having a shave. Remembers wallpaper with dolls.[*]

2
Seeing a very dark man at our gate. I was standing by a high wooden gate. The gate opened and I saw the darkest man I've ever seen: skin like bark, thick black stubble, large black eyes. When he spoke I couldn't understand him,

---

[*] Vicky Swindales, quoted in BBC4 radio programme, *In My Pram I Remember*, Wednesday, 24 January 2007.

and I ran to my mother. She gave him something. Money perhaps. He went. I cried. He was an Italian ex-prisoner of war who was looking for work.

2

1. Playing with dog and cat in farmyard.
2. Camp beds with grey blankets on which we had to 'lie down and rest' during first few days in school.

2

Rocking-horse and moving pedal-operated toy horse in large playroom.

2

On a horse I am very tall, it felt wonderful.

2

My earliest memory is of me standing on my Grandpa Alf's knees, whilst he was seated. He had a almost completely bald head that was smooth and shiny and I remember putchkering (sp? Yiddish word—loose translation, kneading) with his head perhaps because it was such a strange thing for a young child to behold. (My other relatives had hair!)

2

Playing with dog and cat in farmyard. First holiday: staying with parents on working farm in Cardiganshire. There are photographs taken around the farm, including me and a dog.

2

My brother walked me down our terraced street in Liverpool. I held Saville's hand. We crossed over the big road, which was usually busy and full of cars. There were lots of shops and Saville took me to stand next to one that was on a corner. We were standing for ages, just waiting. Then the cars seemed to stop coming down the main road and it was empty for a bit, until a policeman on a motorbike came from my right-hand side and stopped in the road a few feet from where we were standing. I can't remember seeing lots of people, but there seemed to be more noise as the motorbike stopped.

We had to wait even longer before there were any more cars on the main road and next were two more motorbikes with policemen on and right behind was a black car. I could see the people sitting in the back facing each other like they would on a bus. The back window of the car was down and one of the men had his head out of it a bit and was waving. He had black hair and it was bushy. I waved back. The motorbikes and the car were driving quite fast and they passed me and drove off along the main road. [This was the Beatles]

2

I was in my pram and I woke up looking at the sun making a line of shadow on the houses opposite us on Rosecroft Avenue. I do remember very clearly the line of the shadow and the sort of beigey look of the walls (they were pebbledash I found out many years later). BUT the really odd thing is that I remember thinking (as a thought-shape not as words of course) 'Here I am again'. So I must have known I'd been there before. I remember being taken out of the pram and being held up to look at my grandmother and my mother in the kitchen.

2

Holding a woman's hand entering an apartment, walking in the narrow entryway and seeing a bathtub on the left.

2

1. Lying looking up at the sky.
2. Sitting on my father's shoulders in a crowd waving a flag.

2

Standing and looking out and down at a sunlit garden and calling out 'Cecil' to a man I could see below. Trying to scoop up the water from a birdbath with a metal jug.

2

Lying down and looking up at a blue sky with a terracotta red tiled roof and hearing and seeing a piston-engined aircraft flying slowly overhead.

2

1. Perhaps my earliest memory was of being taken to the doctor in Nairobi for an injection. His name was Dr Trim. We had to mount a flight of stairs, then turn right at the top of the stairs through a door to see the doctor.
2. Around the same time, I remember a dream. In my dream I had two mothers. One was a lost mother, lost from earliest babyhood, with a brown face, and long dark hair which hung down and enveloped me as she bent over my bed and hugged me. The other had shorter hair and was less 'motherly'. Both were my mother, but the long-haired one was my true mother from long ago, now lost and dimly remembered. This dream does not reflect anything unhappy.

2.33

I am watching an animated figure of a man in the window of a tall tower. He blows a musical instrument and red tulips come up before him. He finishes and he goes back inside the tower. Two shutters close. Moments later he comes out of the side of the tower and floats on a magic carpet across to a second tower. He goes in the side of this second tower and a moment later two shutters open on the front of this tower and I can see him in the window. Again he blows his flute / trumpet and this time yellow tulips grow in front of him. When the tulips are fully up, the man again goes back into the tower, the shutters close and he comes out of the side of the tower to return to the first tower. The sequence repeats itself over and over.

2.5

I have a clear memory of being taken through a gap in the barbed wire on some sand dunes to get to the beach in Dorset. . . . This little island of memory largely features the sand dunes, the marram grass, and the barbed wire with its gap.

2.5

A giant hole in the ground, with lamp-posts at crazy angles surrounding it, and snow on the ground.

2.5

Playing on a grassy bank.

2.5

Playing the horse race game (the model horses being advanced along a numbered path according to the throw of dice, or some such gambling device) on a ship, while my mother was back in the cabin, suffering from some terrible rash on her face.

2.5

I am sitting in the back of a London taxi. The convertible part of the roof covering the passenger compartment has been lowered so we are open to the air. We are stationary in the midst of a large, noisy but happy crowd which seems to stretch as far as I could see.

2.5

1. Iron front gate.
2. Banging head on metal underground, and crying.
3. Loud bang, confusion everywhere, looking out of upstairs window.

2.5

Lying on my back looking up at the sky with the leaves moving.

2.75

A game: being thrown through the air by large men in a big room—a sweet smell from them. Laughter. My grandparents had a huge house in Birmingham where, at the end of WW2, they put up American servicemen, who played boisterous games and chewed gum.

2.8

Both parents swimming out to sea, becoming invisible from the sandcastle on which I was standing with my brother and sister, and fearing that parents would not be coming back. Comforted by sister.

3

1. I was standing on the balcony of my parents' flat in Park Crescent, watching soldiers marching by. I remember drawing my parents' attention to the way the men were swinging their arms.
2. Staying with my nanny's family in the country. I remember my father standing in a cornfield wearing a light linen jacket and holding a glass of cider in his hands.
3. Lying in the darkness and hearing the sirens. Contentless terror.

3

I am looking down a grating at a golliwog which has dropped down somehow. I am sad. The sun is shining and has a face like the lemon in the 'Idris when I's dry' advertisement for a brand of lemon squash. This advertisement features in another memory which I think is based on what I was told by my mother, which is that when I saw this advertisement on a bus, I said 'Funny lemon a-crying.' I seem to remember that I found the face rather frightening.

3

I am in a bathroom (large, bright, perhaps sumptuous). I have shat my pants, and am trying without much success to clean them in a sink. Feelings of fear at impending punishment.

3

1. A dream. Standing at the edge of my Grandparents' lawn looking down at neat rows of cabbages. The cabbages looked like vegetable faces, and they laughed as I played on the lawn.

2. The first time I met my Father. I stood with my Mother and Grandparents in the little hallway of my Grandparents' house as a big smiling man in uniform came up the steps through the front door to scoop me into his arms. My father went abroad serving in the RAMC shortly before I was born in 1943. He did not return until his surgical duties allowed him home in I think 1946 when I would have been about 3. The night (or a few nights) before his return I had the cabbage dream, which for some reason made a huge impression on me. When we met and he picked me up I think I felt confused, and perhaps alarmed. Although I have no distinct memory of saying it, I have often been told that my first words to my father were 'The cabbages laughed at me.'

3

Playing with my grandfather's porcelain-eye before he put it back into place.

3

Riding a tricycle in Turner Close.

3

1. I am in the rear seat of a car, looking upward through the window, seeing an immensely strange sort of metal web flying by overhead. I feel extremely small, the car feels immense, the impression of the 'web' is of something vast, weird, and frightening. [The strange 'web' was the superstructure of girders of a bridge] However, I am burrowed cosily into the seat of the car, probably (but not certainly) alongside my mother. My father is driving the car. Also, I love being in the car and in motion.

2. I am climbing the long flight of stone steps at the top of the hill near my home. For the first time, I am climbing the stairs by myself. I am thrilled at my power. I am determined to get to the top. My mother is with me close by, to see that I do not fall. I am not afraid, because she is near. There is a tall fir tree at the top, and at Christmastime it is all lit up. I love the tree, and want to see it. Also, from the top of the stairs, I feel that I can see the whole world.

3

1. Being very afraid in bed next to my mother when there was a terrible noise outside.

2. Watching soldiers marching past our house singing.

3

Leaning over the parapet of a bridge across railway tracks. My father is holding me up to watch trains. I am alarmed to be in his care and away from my mother, and dislike the smelly clouds of steam that envelop us as each train roars under the bridge.

3.3

Hiding, afraid, under the kitchen table in my grandmother's house in Merthyr Tydfil when an aeroplane went overhead.

3.3

Sitting (trapped?) in a wooden high chair, fronted with a little table top, with an insect buzzing around me, strong visual memory of a woman flapping it away, possibly with a cloth.

I have another early memory, with no idea of which is first, of being in some spectator stands with my family and seeing a procession, horses, soldiers, coaches, etc. This is associated with a vague recollection of my younger sister wanting to pee and my mother whipping out a plastic pot. The second memory I had always assumed was the 1953 coronation in London until I mentioned it to my mother who says we were actually at the rehearsal for the trooping of the colour that same year, which makes it June 1953.

3.4

Running between two people, whom I keep alternately kissing: one my new baby brother, the other I'm not sure about. But it is someone I'm glad to see because I am torn between these two who are sitting at opposite corners of the living-room in my house. Someone is crying—I think it's my mother; but (for once) I'm not alarmed by this. I feel elated and the centre of attention. There are other people in the room too and, in spite of this definite, isolated area of sadness (or, more specifically, crying), the general feeling is a happy one.

3.5

A cramped, enclosed space, dimly lit. My grandmother, in a dark dress patterned with tiny white flowers, sits at an old green-baize card-table knitting. My two brothers lie stacked one above the other on rough wooden bunks, wrapped in grey blankets. I am in my blue carry-cot.

3.5

Being in my bedroom dressed only in a pair of knickers which I had filled with plastic Noddy / Big Ears figures and my mum being really angry with me.

3.5

A red brick wall, topped with a white corbel, rough to the touch. In back of the wall, a window, and behind the window, my mother.... The wall was at least five feet high, so I must have been held by my father, and placed my hands on the surface of the wall, whose texture remains the most vivid portion of the memory.

3.5

1. I used to imagine that I could remember my father playing with our labrador dog. It was fond imagination based on a photograph. I was only 21 months when he left to fight in Egypt never to return.

2. Standing by our back door watching collie eating from a bowl when it turned on me ... Blood. General panic. Everybody blaming each other, but making a fuss of me. Apparently I had been standing between the dog and its master, a shepherd. It bit me perilously close to the eye. I was 3.5.

3. Going for a walk. Finding a wild damson tree. Gorging myself. Returning home I helped our gardener putting carrots in sand for the winter. Belching made me feel rather sick. I went to the garage and lay down on the back seat of the car. Apparently I passed out and was discovered a few hours later. The doctor was called and gave me morphine. Age about 4.

4. Fearing the Japanese might invade; they sounded more bloodthirsty than the Germans, and burying myself in my bed hoping they might not see me. Age about 4.5.

4

Two memories that seem to be earliest. When we travelled out of suburban London, we used to stop at a place with a blue fountain, next to a river crossed by a wooden footbridge. On one occasion, the river had risen, and was flowing over the bridge: I remember the odd sensation of seeing water above the bridge, as it were, flowing over the wooden decking: and the image disturbs me to this day. I believe we were allowed to walk across the bridge: it was both fascinating and frightening. Presumably not that dangerous!

Travelling with my mother by train to Eastbourne where my grandfather lived. He was a Presbyterian minister and lived in a rather wonderful Manse, full of mysterious rooms and long corridors. I remember the series of tunnels on the railway through Sussex. Then I am in the corridor outside the 'pantry' in the Manse. This was at the end of a corridor leading to the very large gothic front door. I remember my father turning up and ringing or knocking at the front door. I was at the back of the house for a reason—and I have wondered if there was an issue about letting my father in . . . No resolution.

4

It was springtime. I was four. We were moving to a new town. We were moving into a new house. I wasn't too sure about all of this. I know I was four at the time because my mother told me so. We arrived at the new house. It was very noisy and crowded at the new house and no one was paying attention to me. I decided to go for a walk. The day was sunny and warm and it felt good. Then I got lost. I couldn't find my way back. I became convinced my family had left me behind and gone back to the old house—my house. I started to cry. A policecar drove up and they gently convinced me to go with them. I sat between the two policemen in the front of the car. The policeman to my right shared a stick of gum with me. I stopped crying. We circled around the block and went back to the 'new house'. My mother, father, siblings, and my Aunt Peggy all came out to greet us. I had no idea why they were so angry. After all, they were the ones who abandoned me!

4

Seeing from the butcher's shop, where my mother was making a purchase, a sports car screech to a halt at a zebra crossing but not in time to prevent an elderly woman being knocked to the ground and Bill Drogan (I remember his name) the butcher's assistant running out with bloody hands (from the meat!) to assist the woman.

4.25

Animals. A rabbit in its kennel (it was actually more like a bird's cage) and many many cats. Fragments of a scene in which the two—the cat and the rabbit who was in our yard met and the disastrous result of the meeting.

4.25

My father picking a fruit from a tree in our garden, and telling me its name: 'greengage'. He could reach the fruit easily because he was so tall.

Being put to sit on the end of my baby sister's pram, facing towards our nanny, during a walk. Autumn leaves on the ground.

Being hit (accidentally) on the head by a sharp crescent-shaped garden tool, with which my brother was playing, when I was standing near him. Having it bathed by my mother with cotton wool and water.

4.5

Every morning we would dash down the drive. Halfway along we crept along the strip of grass at the side and carefully pulled away the branches and bracken that camouflaged a large hole. Nothing was ever there. We, my uncle and I, had dug a huge pit—it was our war effort. If the Germans invaded and marched up the drive to capture us, they would fall into our pit and we would be saved. We called it 'the Hitler trap'.

4.5

I remember always wanting a puppy, so when I got him, he was the centre of my young universe. I called him Max. One day, not long after I got him, we were playing in the garden when I decided to ask a friend from across the road to come and play. I checked with my mum who said that this was OK but to shut the gate behind me so that Max did not get out. So I went back out into the garden and deliberately left the gate open so he could follow me. I remember vividly thinking that this was wrong and briefly deliberating over it, but I really wanted him to come.

As I walked up the driveway of my friend's house, I heard a horrible noise— this must have been the brakes screeching, I ran down the driveway and saw a red-haired man, bending over Max who was in the gutter. I recognized him instantly as Mr Grundy, a man who lived in our road with his family of ('strange' or so my friends and I thought) red-haired children. I ran over to Max and knelt down beside him. I was crying so hard I couldn't see properly. Max was making a horrible sound and when I put my hand on him, he tried to bite me, which made me cry even more. I remember thinking he hated me for leaving the gate open and could see it was my fault. Through my tears I remember watching his red blood trickling down the drain that he was lying beside, then my dad came out and took me inside. The next thing I remember is dad coming in as we were sitting down to supper and telling us Max had died. I felt so guilty about this for a long time, and remember hating Mr Grundy for running over my puppy.

4.5

I remember being told I would soon be starting school, but not understanding what this meant. In my puzzlement all I could do was associate the sound of the word 'school' with that of 'nail'. I knew what a nail was. Clearly this didn't get me very far.

5

The sound of a grandmother clock chiming.

5

Nursery school, 1950s, after lunch getting out mattresses and cots, normally stacked at the side of the room, then lying down for a rest and having stories read to us. The only book - and there must have been many—that I remember very clearly is John Bunyan's Pilgrims Progress—frightening and extremely dark. The image of myself as a child lying down and hearing about Christian's travails has often come back to me in later life in different ways especially the eagerness with which I longed for the next instalment: did he survive?

5

One early memory is of my father getting so soaked on his way back from the bus that he had to take all his clothes off in the living-room. He leapt around laughing, dangling all over the place, which my mother and I found very funny and also, in my case, rather rude. But there occurred to me at the time the image of him walking down the path at the side of the house to the bus stop a few hours earlier, when it was such a nice, sunny day that he hadn't bothered to take a mac or umbrella. It seemed a somewhat cruel contrast with the ferocious rain that had come out of nowhere and I saw it as a warning that things can change rapidly. (It would not be my earliest memory, as I can remember the day when my brother was born, in the August before my fifth birthday.)

5

I stand there among the trees in snow and see my mother's back in a dark fur coat running away from me. I feel surprise with the funny womanly way she runs. I wonder whether I will be able to run that way.

5

On toboggan in snow.

5

A television in the corner of the room. The screen is black and white. There is a coffin and some people. [Churchill's funeral.]

6

My earliest memory is of my mother and me sitting on a box in a room full of boxes but with no furniture in our new house when I was six years of age.

7

I remember a boy with red hair and freckles eating limpets with a jack-knife, sitting near the end of a stone jetty. I felt very sick.

Notes

1 Richard J. McNally, *Remembering Trauma* (The Belknap Press of Harvard University Press, 2005), 43–4.

2 Charles Darwin, *The Correspondence of Charles Darwin*, Vol. 2 (1837–43), Appendix III, 'An Autobiographical Fragment'.

3 BBC Radio 4, *In my pram I remember*, 11 a.m., 24 January 2007.

4 Quoted in David B. Pillemer, *Momentous Events, Vivid Memories* (Harvard University Press, 1998), 45.

5 Donald P. Spence, 'Passive remembering', chapter in Ulric Neisser and Eugene Winograd, *Remembering Reconsidered* (Cambridge University Press, 1988), 312.

6 Daniel L. Schacter, *How the Mind Forgets and Remembers* (Souvenir Press, 2003), 127.

7 Christian therapist 'Sam Holde', quoted in Mark Pendergrast, *Victims of Memory* (HarperCollins, 1996), 201–2.

8 M. J. Eacott and R. A. Crawley, 'The offset of childhood amnesia', *Journal of Experimental Psychology*: General, 1998, Vol. 127, No. 1, 22–33.

9 Ibid.

10 Ibid.

11 S. Freud, 'Introductory lectures on psycho-analysis', in J. Strachey (ed.), *The Standard Edition of the Complete Psychological Works of Sigmund Freud*, vii. 326.

12 Yasushi Miyashita, 'Cognitive memory: cellular and network machineries and their top–down control', *Science* 306 (15 October 2004), 435–40.

13 http://en.wikipedia.org/wiki/Frontal_lobe

14 Susan A. Clancy, *Abducted: How People Come to Believe they were Kidnapped by Aliens* (Harvard University Press, 2005), 69.

15 Pillemer, *Momentous Events, Vivid Memories*, 133.

16 Ibid. 130.

17 Gabrielle Simcock and Harlene Hayne, 'Breaking the barrier?', *Psychological Science*, Vol. 13, No. 3 (May 2002), 229.

18 Rick Richardson and Harlene Hayne, 'You can't take it with you', *Current Directions in Psychological Science*, Vol. 16, No. 4 (August 2007), p. 225.

19 Schacter, *How the Mind Forgets and Remembers*, 126.

20 Ruth Garner, Mark G. Gillingham, and C. Stephen White, 'Effects of "seductive details" on macroprocessing and microprocessing in adults and children', *Cognition and Instruction*, Vol. 6, No. 1 (1989), 41–57.

21 Pillemer, *Momentous Events, Vivid Memories*, 81.

22 P. J. Bauer, J. A. Hertsgaard, and G. A. Dow, 'After 8 months have passed—long-term recall of events by 1-year-old to 2-year-old children', *Memory* 2 (1994), 353–82.

23 Patrick Bateson and Paul Martin, *Design for a Life* (Simon and Schuster, 2000), 168.

24 Katherine Nelson, 'The ontogeny of memory for real events', chapter in Ulric Neisser and Eugene Winograd, *Remembering Reconsidered* (Cambridge University Press, 1988), 269.

25 Ibid. 273.

26 Ibid. 265.

27 Ibid.

28 Emily Oster, Foreword, in Katherine Nelson (ed.), *Narratives from the Crib* (Harvard University Press, 2006), p. v.

29 Nelson, 'The ontogeny of memory for real events', 266–7.

30 Quoted in Pillemer, *Momentous Events, Vivid Memories*, 122–3.

31 McNally, *Remembering Trauma*, 35.

32 Qi Wang, 'Earliest recollections of self and others in European American and Taiwanese young adults, *Psychological Science*, Vol. 17, No. 8 (August 2006), p. 712.

33 Ibid. 713.

34 Jens Brockmeier and Qi Wang, 'Where does my past begin? Lessons from recent cross-cultural studies of autobiographical memory', Third Conference for Sociocultural Research, www.fae.unicamp.br/br2000/trabs/2550.doc

35 Kate E. Fiske and David B. Pillemer, 'Adult recollections of earliest childhood dreams: a cross-cultural study', *Memory* 14/1 (2006), 57–67.

36 Pillemer, *Momentous Events, Vivid Memories*, 199.

37 Schacter, *How the Mind Forgets and Remembers*, 3.

38 Socrates to Theaetetus. Plato, *Theaetetus* 191d, quoted on http://mnemosynosis.livejournal.com/profile

39 Frederic Bartlett, *Remembering: A Study in Experimental and Social Psychology* (Cambridge University Press, 1932, 1995), 75.

40 Ibid. 166.

41 Ibid. 150.

42 Ibid. 3.

43 Ibid.

44 Ibid. 33.

45 Eric Otto Siepmann, *Confessions of a Nihilist* (Victor Gollancz, 1955), 177–8.

46 Bartlett, *Remembering*, 44.

47 F. C. Bartlett, *The Mind at Work and Play* (George Allen & Unwin; American edition: Boston, Beacon Press, 1951).

48 Charles P. Thompson, John J. Skowronski, Steen F, Larsen, and Andrew Betz, *Autobiographical Memory: Remembering What and Remembering When* (Lawrence Erlbaum Associates, 1996), 138.

49 Roger C. Schank and Robert P. Abelson, in Robert S. Wyer, Jr (ed.), *Knowledge and Memory: The Real Story* (Lawrence Erlbaum Associates, 1995), 1–85.

50 F. C. Bartlett, 'Notes on Remembering' (1960), http://www.bartlett.sps.cam.ac.uk/NotesOnRemembering.htm

51 Schacter, *How the Mind Forgets and Remembers*, 82.

52 Richard McNally, 'Betrayal trauma theory: a critical appraisal', *Memory*, 15/3, 280–94.

53 Ibid.

54 J. Don Read and D. Stephen Lindsay, *Journal of Traumatic Stress*, Vol. 13, No. 1 (2000), 144–5.

55 Harry N. MacLean, *Once Upon a Time* (HarperCollins, 1993), 349.

56 Ibid.

57 Elizabeth F. Loftus, 'Autobiography', in G. Lindzey and M. Runyan (eds.), *History of Psychology in Autobiography*, Vol. IX (American Psychological Association Press, 2007), 198–227.

58 Ibid.

59 Quoted in Pendergrast, *Victims of Memory*, note, pp. 438–9.

60 Quoted ibid. 440–1.

61 McNally, *Remembering Trauma*, 7.

62 Recovered memory patient on computer bulletin board, quoted in Pendergrast, *Victims of Memory*, 1.

63 Quoted ibid. 258.

64 Quoted ibid. 351.

65 Quoted ibid. 367.

66 Quoted ibid. 276.

67 British consultant psychiatrist, quoted ibid. 223.

68 Ibid. 258.

69 Quoted ibid. 357.

70 Quoted ibid. 358.

71 McNally, *Remembering Trauma*, 174.

72 Ibid. 192.

73 Richard A. Gardner, Institute for Psychological Therapies, http://www.iptforensics.com/journal/volume4/j4_4_3.htm

74 Ibid.

75 McNally, *Remembering Trauma*, 151.

76 Did you notice that during the game a man dressed in a gorilla suit walked from left to right, stopping for a moment to wave at the camera? Most people concentrating on the task don't see the man in the gorilla suit. Not mentioning

the gorilla when describing this scene later is not a failure of memory but of encoding.

77 McNally, *Remembering* Trauma, 176–7.

78 J. Don Read and D. Stephen Lindsay, ' "Amnesia" for summer camps and high school graduation', *Journal of Traumatic Stress*, Vol. 13, No. 1 (2000), 129–47.

79 http://www.bbsonline.org/Preprints/Erdelyi-04022004/Referees/

80 Harlene Hayne, Maryanne Garry, and Elizabeth Loftus, comments on M. H. Erdelyi, *What is Repression?* (2006), 'The unified theory of repression', *Behavioral and Brain Sciences* (29), 521.

81 Jennifer J. Freyd, 'Science in the memory debate', *Ethics and Behavior*, 8/2 (1998), 101–13.

82 Clancy, *Abducted*, 69.

83 *New York Review of Books*, Vol. 53, No. 11 (22 June 2006).

84 Jennifer J. Freyd and David H. Gleaves, *Journal of Experimental Psychology*, Vol. 22, No. 3 (1996), 811–13.

85 J. J. Freyd, S. R. Martorello, J. S. Alvarado, A. E. Hayes, and H. C. Christman, 'Cognitive environments and dissociative tendencies: performance on the standard Stroop task for high versus low dissociators', *Applied Cognitive Psychology*, 12 (1998), S91–S103.

86 Loftus, 'Autobiography'.

87 Douwe Draaisma, *Why Life Speeds Up as You Get Older: How Memory Shapes our Past* (Cambridge University Press, 2004), 24.

88 Loftus, 'Autobiography'.

89 Pendergrast, *Victims of Memory*, 407.

90 http://www.law.umkc.edu/faculty/projects/ftrials/mcmartin/mcmartinaccount.html

91 Lona Manning, *Nightmare at the Day Care: The Wee Care Case*, http://www.crimemagazine.com/daycare.htm

92 M. Bruck *et al.*, *Developmental Review* 22 (2002), 520–54.

93 Ibid.

94 Ibid.

95 McNally, *Remembering Trauma*, 273.

96 Clancy, *Abducted*, 30–1.

97 Anonymous Californian past-life hypnotherapist, quoted in Pendergrast, *Victims of Memory*, 237.

98 Bryan, *Close Encounters*, 419, quoted ibid. 104.

99 Clancy, *Abducted*, 33.

100 *http://www.bobbyjordanfansite.com/articles.htm*

101 Pillemer, *Momentous Events, Vivid Memories*, 146.

102 Alice Miller, *Breaking Down the Wall of Silence* (Virago Press, 1991), quoted in Pendergrast, *Victims of Memory*, note, p. 367.

103 Quoted ibid. 323.

104 Incest therapist 'Janet Griffin', quoted ibid. 206–7.

105 Quoted in *Journal of Consulting and Clinical Psychology*, Vol. 62, No. 6 (1994), 1177–81.

106 Mike Lew, *Victims No Longer*, quoted in Pendergrast, *Victims of Memory*, 34.

107 Quote from Renee Fredrickson, *Repressed Memories*, in Pendergrast, *Victims of Memory*, 37.

108 US therapist 'Leslie Watkins', quoted ibid. 214.

109 US therapist 'Leslie Watkins', quoted ibid. 215.

110 US therapist 'Leslie Watkins', quoted ibid.

111 UK therapist 'Delia Wadsworth', quoted ibid. 219–20.

112 US retractor therapist, quoted ibid. 245.

113 British consultant psychiatrist, quoted ibid. 226.

114 Ibid. 170.

115 Janet Walker, 'The traumatic paradox: documentary films, historical fictions, and cataclysmic past events', *Signs*, Vol. 22, No. 4 (Summer 1997), 803–25.

116 McNally, *Remembering Trauma*, 21.

117 Quoted in Pendergrast, *Victims of Memory*, 269.

118 Quoted ibid. 356.

119 Quoted ibid. 320.

120 McNally, *Remembering Trauma*, 116.

121 Steven Jay Lynn, 'Memory recovery techniques in psychotherapy: problems and pitfalls', *Skeptical Inquirer*, July/August 2003.

122 Quoted in Pendergrast, *Victims of Memory*, 273.

123 Quoted ibid. 377.

124 Clancy, *Abducted*, 30–1.

125 Ibid.

126 Quoted in Pendergrast, *Victims of Memory*, 273.

127 Quoted ibid. 377.

128 Clancy, *Abducted*, 50.

129 McNally, *Remembering Trauma*, 74.

130 'Survivor' Betty Peterson, quoted in Pendergrast, *Victims of Memory*, 58.

131 McNally, *Remembering Trauma*, 248.

132 Hamish Pitceathly, British psychotherapist, quoted in Pendergrast, *Victims of Memory*, 228.

133 Incest therapist 'Janet Griffin', quoted ibid. 208.

134 Quoted ibid. 357.

135 US retractor therapist, quoted ibid. 243.

136 Quoted ibid. 359.

137 Walker, 'The traumatic paradox'.

138 Maryanne Garry, Elizabeth F. Loftus, Scott W. Brown, and Susan C. DuBreuil, 'Womb with a view: memory beliefs and memory-work experiences', chapter in David G. Payne and Frederick G. Conrad (eds.), *Intersections in Basic and Applied Memory Research* (Lawrence Erlbaum Associates, 1997).

139 Anonymous Californian past-life hypnotherapist, quoted in Pendergrast, *Victims of Memory*, 234.

140 Hamish Pitceathly, British psychotherapist, quoted in Pendergrast, *Victims of Memory*, 231.

141 Melvin Harris, *Sorry, You've Been Duped* (Weidenfeld & Nicolson, 1986), 160–2.

142 Anonymous Californian past-life hypnotherapist, quoted in Pendergrast, *Victims of Memory*, 235.

143 John F. Kihlstrom, 'Hypnosis, memory and amnesia', *Phil. Trans. R. Soc. Lond.* B (1997), 352, 1727.

144 Ibid, 1731.

145 Ibid.

146 http://www.smith-lawfirm.com/Murphy_Memory_Article.html#14.

147 Clarissa Dickson Wright, *Spilling the Beans* (Hodder & Stoughton, 2007), 27–8.

148 B. Rind, P. Tromovitch, and R. Bauserman, 'A meta-analytic examination of assumed properties of child sexual abuse using college samples', *Psychological Bulletin* 124 (1998), 22–53.

149 Ibid.

150 McNally, *Remembering Trauma*, 94–5.

151 Loftus, 'Autobiography'.

152 http://www.casp.net/taus-1.html

153 Loftus, 'Autobiography'.

154 Pendergrast, *Victims of Memory*, 222.

155 McNally, *Remembering Trauma*, 42.

156 BFMS Newsletter, Vol. 12, No. 1.

157 Pendergrast, *Victims of Memory*, p. xxv.

158 http://everything2.com/index.pl?node_id=1541468

159 US retractor therapist, quoted in Pendergrast, *Victims of Memory*, 249–50.

160 Quoted ibid. 371.

161 Quoted ibid. 391.

162 *Institute for Psychological Therapies*, Vol. 3 (1991), www.ipt-forensics.com/journal/volume3/j3_3_3.htm

163 Mike Stanton, 'U-turn on memory lane', *Columbia Journalism Review*, July/August 1997.

164 Pendergrast, *Victims of Memory*, 56.

165 Pillemer, *Momentous Events, Vivid Memories*, 10.

166 British consultant psychiatrist, quoted in Pendergrast, *Victims of Memory*, 222.

167 Incest therapist 'Janet Griffin', quoted ibid. 208.

168 Quoted ibid. 314.
169 *http://www.oxford.anglican.org/files/stem/PROTECTING_ALL_GOD_copy.doc. pdf* (When accessed in October 2008, this publication still included *The Courage to Heal* among its list of 'useful books'.)
170 Norman Brand (ed.), *Fractured Families* (BFMS, 2007).
171 Ibid.
172 Ellen Bass and Laura Davis, *The Courage to Heal* (Vermilion, 1988), 22.
173 http://www.studentaffairs.cmu.edu/counseling/concerns/child.html
174 http://www.counselingcenter.uiuc.edu/?page_id=170
175 Quoted in Pendergrast, *Victims of Memory*, 359.

Index